Process and Reaction Flavors

ACS SYMPOSIUM SERIES **905**

Process and Reaction Flavors

Recent Developments

Deepthi K. Weerasinghe, Editor
The Coca-Cola Company

Mathias K. Sucan, Editor
Quest International

**Sponsored by the ACS Division of
Agricultural and Food Chemistry and
ACS Corporate Associates**

American Chemical Society, Washington, DC

Library of Congress Cataloging-in-Publication Data

Process and reaction flavors : recent developments / Deepthi K. Weerasinghe, editor, Mathias K. Sucan, editor ; sponsored by the ACS Division of Agricultural and Food Chemistry

p. cm.—(ACS symposium series ; 905)

Includes bibliographical references and index.

ISBN 0–8412–3905–3 (alk. paper)

1. Maillard reaction—Congresses. 2. Flavor—Congresses. 3. Food industry and trade—Safety measures.

I. Weerasinghe, Deepthi K., 1953- II. Sucan, Mathias K. III. American Chemical Society. Division of Agricultural and Food Chemistry. IV. Series.

TP372.55.M35P76   2005
664′.5—dc22   2005041110

The paper used in this publication meets the minimum requirements of American National Standard for Information Sciences—Permanence of Paper for Printed Library Materials, ANSI Z39.48–1984.

PRINTED IN THE UNITED STATES OF AMERICA

Foreword

The ACS Symposium Series was first published in 1974 to provide a mechanism for publishing symposia quickly in book form. The purpose of the series is to publish timely, comprehensive books developed from ACS sponsored symposia based on current scientific research. Occasionally, books are developed from symposia sponsored by other organizations when the topic is of keen interest to the chemistry audience.

Before agreeing to publish a book, the proposed table of contents is reviewed for appropriate and comprehensive coverage and for interest to the audience. Some papers may be excluded to better focus the book; others may be added to provide comprehensiveness. When appropriate, overview or introductory chapters are added. Drafts of chapters are peer-reviewed prior to final acceptance or rejection, and manuscripts are prepared in camera-ready format.

As a rule, only original research papers and original review papers are included in the volumes. Verbatim reproductions of previously published papers are not accepted.

ACS Books Department

Contents

Process Flavors from Classical Maillard Reactions

Analysis of Process Flavors

Preface

Process flavors have come to stay as an important as well as a cost effective method of producing complex flavors. Progress in the understanding a nd utilization of process flavors was made due to new technology; regulation that meet consumer safety concerns; and the industry demand for better, complex, and authentic products. The flavor industry is by far the largest user of knowledge from process- and reaction-flavor studies and has grown from 300–350 millions 10 years ago to more than 10 billions today.

Maillard r eaction; lipid oxidation and degradation; caramelization; degradation of sugars, proteins, and vitamins; and the interactions of degradation products are the chemical platform for generating many flavor compounds encountered in processed flavorings, flavors and foods. Clearly, there is growing acceptance of these flavors in the world food supply,

This symposium was organized to shed some light on the current state of science in process and reaction flavors and to report recent significant findings. The book provides a comprehensive overview of process a nd r eaction f lavors t hat i s f ollowed b y a d iscusion o f safety, legal, and regulatory aspects, including an i ntroduction t o K osher a nd Halal issues. In the next section ingredients and intermediate Maillard reactions are discussed. The section on classical Maillard reaction discusses some new methodology in the formation of reaction flavors and the final section discusses the analytical challenges that are faced by chemists with respect to the identification and monitoring of flavor molecules in process flavors.

We a re m ost g rateful t o t he d iverse group of authors for their outstanding effort in the preparation of this book. This book is intended to be a reference book for researches both in the industry and the academia.

Deepthi K Weerasinghe
The Coca-Cola Company
P.O. Box 1734
Atlanta, GA 30301

Mathias A. Sucan
Quest International
5115 Sedge Boulevard
Hoffman Estates, IL 60192

Process and Reaction Flavors

Chapter 1

Process and Reaction Flavors: An Overview

Mathias K. Sucan[1] and Deepthi K. Weerasinghe[2,3]

[1]Quest International, 5115 Sedge Boulevard, Hoffman Estates, IL 60192
[2]dP3 Consulting, 8 Madison Drive, Plainsboro, NJ 08536
[3]Current address: The Coca-Cola Company, P.O. Box 1734, Atlanta, GA 30301

Maillard reaction, lipid oxidation, degradation of sugars, proteins, lipids, ribonucleotides, pigments and vitamins, and the interactions of degradation products are the chemical platform for generating many flavor compounds encountered in process and reaction flavors, flavorings and foods. The flavor industry is by far the largest user of knowledge from process/and reaction flavor studies and has grown from approximately 2 billions, 20 years ago, to about 8 billion dollars in annual sales today. During the last few decades, much progress in the understanding and utilization of process flavors was made due to advances in chromatographic separation and computer-related technology, relentless investigation of a wide range of flavor precursors, regulation that met consumer safety concerns, and industry demand for better, complex and authentic products. This symposium was organized to shed some light on the current state of science in process and reaction flavors, and to report recent significant findings.

Flavor Industry and Flavor Technology

The global market for the flavor and fragrance industry is estimated at more than 15 billion dollars today (1,2). Approximately 525 companies contribute to this market. About half of the 15.1 billion dollar worldwide flavor & fragrance sales is flavor, which splits between dairy (14%), savory (23%), beverages (31%) and other (32%). The market has almost quadrupled over the last 20 years (figure 1). However, the last few years registered slow growth, which has resulted in mergers and acquisitions, downsizing and cutback on innovation initiatives.

Restructuring has lead to 10 companies controlling about 2/3 of the market. James Giese (3) recently reported that in 1995, 10 companies made up 59% of the market share, but by 2000, 8 companies had 71% of the market share. The drawback of all these mergers and acquisitions is the elimination of technical positions and the reduction of R&D budgets. A recent report of the number of patents in process and reaction flavors filed during the last quarter of century (4) indicates a steady increase from 1980 to 2000, then a steep decline thereafter (figure 1), which may be due to a significant reduction in R&D resources. The trends in patent generation also find good correlation with flavor sales. Ironically, the flavor industry is by far the largest user of knowledge from flavor research, and much progress in the understanding and utilization of process flavors was made due to advances in chromatographic separation and computer-related technology, relentless investigation of a wide range of flavor precursors, regulation that meets consumer safety concerns, and industry demand for better, complex and authentic products. This offering provides background information on the state of knowledge in the aforementioned areas and concludes on key research gaps.

Advances in Chromatographic Separation and Computer-Related Technology

Prior to gas chromatograph (GC), the characterization of unknown volatile flavor compounds was a tedious task. Great strides in flavors research have been made since its invention by James and Martin in 1952. Earlier GC column were packed columns, and by 1960, only 500 flavor compounds had been identified in

3

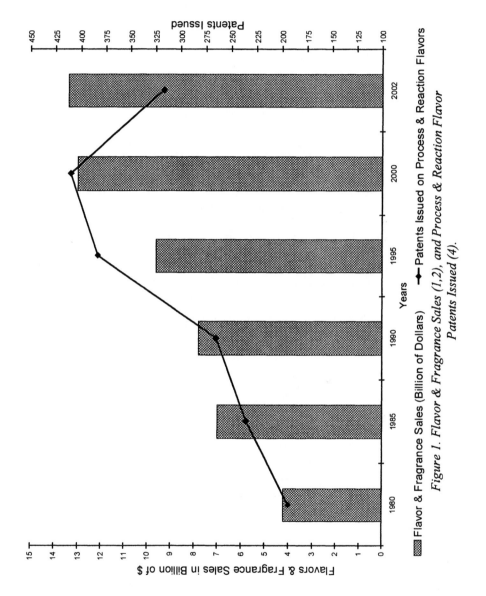

Figure 1. Flavor & Fragrance Sales (1,2), and Process & Reaction Flavor
Patents Issued (4).

foods. The development of high resolution glass capillary columns and tandem gas chromatography/mass spectrometer techniques in the mid 1960's has made possible the identification of tremendous number of flavor constituents. Improvement in GC detectors hardware and progress in computer technologies, have leveraged the magnitude of sensitivity obtainable in the identification of flavor compounds. Improvement in flavor isolation methods was also key contributor to the increased sensitivity. Flavors compounds must be isolated from the complex matrix of foods material before chromatographic analyses. Earlier methods of isolation were sometimes fraught with pitfalls and difficulties, and often sources of artifact formation. Although tremendous progress has been made in terms of limiting artifact formation, real-time aroma release measurement in the mouth, developed by Taylor and co-workers, is a significant development in flavor research (5).

As gas chromatography evolved in sophistication, progress was made in the characterization of flavor compounds and in the elucidation of their chemical pathways. Initial researches focused on the identification of long list of flavor compounds, which has resulted in the identification of more than 7,000 aroma compounds in foods (5). Because, it was impossible to recreate food flavors from the laundry lists of chemicals, earlier attempts were made to determine character impact compounds. Except for few foods, the flavors of many foods are not determined by character impact compounds, rather results from a combination of several key flavor constituents. Today, there are several established methods (CHARM, AEDA, Odor Unit Value) used in the determination of key aroma constituents in foods.

Regulation That Meets Consumer Safety Concerns

Under FEMA (Flavor and Extract Manufacturer's Association) and IOFI (International Organization of Flavor Industry), the flavor industry established an independent expert panel for the review of flavor ingredient safety, certainly because of consumers' awareness and concerns. FEMA introduced a GRAS list of allowed substances in 1965, and has added nearly 2000 compounds since then (6). IOFI is in the process of establishing a list of flavor ingredients it considers safe for use in foods.

FEMA was founded in 1909, and is the sole national association of the flavor industry. FEMA assists in the enactment and enforcement of laws which deal with the rights of flavor manufacturers and consumers, and in the processes which assure its members a supply of safe flavor materials. FEMA's most important activities include safety evaluation of flavor ingredients, protection of member company's intellectual property, establishment of GRAS list, responding to government regulations affecting members, participating in the development and review of flavor ingredient specifications and international regulations of method of safety evaluation, organizing annual membership meeting and monitoring and addressing issues related to flavor labeling (7).

Industry Demand for Better, Complex and Authentic Products

Consumers' demand for authentic and home-like meal flavors prompted the flavor industry to hire corporate chefs to work side-by-side with flavor chemists in the process of flavor development. Flavors from traditional cooking techniques, that take several hours to several days to develop desired profiles, are now matched by flavor chemists. In this process, the corporate chef designs a gold standard food product, and the flavor chemist provides a match to the flavor of the gold standard.

Study of Flavor Precursors

Many compounds encountered in process and reaction flavors derive from Maillard reactions, lipid oxidation, degradation of sugars, proteins, lipids, ribonucleotides, pigments and vitamins, and from the interactions of degradation products. The Maillard reaction is actually a complex group of hundreds of possible reactions. In foodstuffs, Maillard reaction or nonenzymatic browning reaction is responsible for changes in aroma, taste, color and nutritive value during processing and storage. The Maillard reaction in foods has been subject to several reviews (8,9,10,11). An important reaction associated with Maillard reaction is Strecker degradation which involves oxidative deamination and decarboxylation of amino acids in the presence of dicarbonyls. Strecker degradation leads to the formation of aldehydes containing one fewer carbon atom than the original amino acids, and alpha-amino ketones precursors of pyrazines (12). Hydrogen sulfide, ammonia, and acetaldehyde are also formed from the breakdown of mercaptoiminol Strecker intermediate from cystein specific reaction (12,13).

Historical Perspective

The earliest reported scientific studies of Maillard reactions were by Dr. Louis Camille Maillard in figure 2 (14) who, in an attempt to determine the biological synthesis of proteins, heated concentrated solutions of D-glucose and amino

6

acids, and observed a gradual darkening and frothing accompanied by odors reminiscent of the baking of bread and roasting of animal or vegetable products. The elucidation of the chemical pathways in Maillard reaction started with the work of John Edward Hodge (figure 3), of USDA Northern Regional Research Center in Peoria, IL, USA, on the "chemistry of browning reactions". Hodge gained much attention in 1979 among carbohydrate chemists when his 1953 article on "Dehydrated Foods, Chemistry of Browning Reactions in Model Systems", Journal of Agricultural and Food Chemistry, volume 1, issue 15, page 928, was named a "Citation Classic" by the Science Citation Index. It is interesting to note that the mechanism proposed by Hodge in 1953 still provides the basis for our understanding of the early stages of Maillard reactions (12).

Figure 2. Photograph of Professor Louis-Camille Maillard. Courtesy of ACS Symposium Series 215.

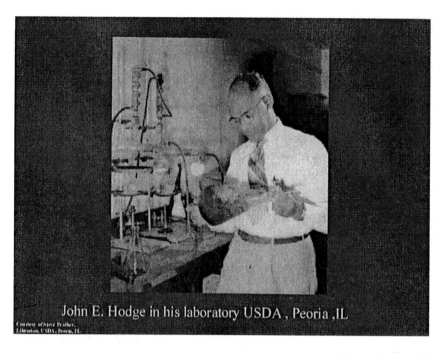

Figure 3. Photograph of Professor John E. Hodge. Courtesy of Steve Prather, Librarian, USDA, Peoria, IL.

The Maillard Reaction

The first step of Maillard reaction series involves Schiff base formation between the carbonyl group of a reducing sugar and the free amino group of an amino acid, peptide or protein. These Schiff bases (also known as N-glucosylamines or N-fructosylamines) can rearrange to form 1-amino-1-deoxy-2-ketose (Amadori compounds) and 1-amino-2-deoxy-2-aldose (Heyns compounds). Amadori and Heyns compounds can undergo dehydration to form furfural (from pentose) or hydroxymethyl furfural (from hexose), or can rearrange into 1-deoxyglycosones or 3-deoxyglycosones (reactive reductones and dehydroreductones). The latter can undergo retro-aldolization reaction to form alpha-dicarbonyls (diacetyl, pyruvaldehyde, acetoin,) which subsequently interact with ammonia and hydrogen sulfide to produce important flavor compounds. Potent odorants formed from Maillard reactions include 3-Methylbutanal, 2,3-butanedione, methional, phenylacetaldehyde, 2-furfurylthiol, 2-ethyl-3,5-methylpyrazine, 2,5-dimethyl-4-hydroxy-3(2H)-furanone, etc. Major classes of Maillard reaction aroma compounds are Strecker aldehydes, and more importantly heterocyclic compounds containing nitrogen, sulfur, oxygen, and combination of these atoms. Maillard reaction products can be specific to the type of amino acids used in the reaction system.

Maillard Specific Reactions

Cystein-specific Maillard reaction include mercaptoacetaldehyde, mercaptopropionic acid and derivatives of thiane, thiolane, thiazine, and thiophene. Proline–specific compounds are pyrrolidins, piperidins, pyrrolizines and azepines. Of methionine-specific compounds, heterocyclic compounds containing a methylthioethyl or methylthiopropyl side chain such as 2-(2-methylthioethyl)-4,5-dimethyloxazole and 2-(3-methylthiopropyl)-5,6-dimethylpyrazine are most interesting. Additionally, methional, methanethiol and 2-propenal are important methionine-specific compounds. Methional can decompose into methanethiol, dimethyl sulfide, dimethyl disulfide and propenal. 3-Methyl-2(1H)-pyrazinone is specific product formed from the reaction of asparagine and glucose. The reaction of glucose with histidine led to the formation of amino acid specific compounds 2-acetyl- and 2-propionyl-pyrido[3,4-d]imidazole along with their tetrahydropyrido derivatives (15).

Thiamine Degradation

The thermal degradation of thiamine is very important in process flavors and has been investigated by several authors (16,17,18,19). Aqueous system at pH 2.3/135 C/30 min generated more decomposition products (carbonyls, furanoids, thiophenoids, thiazoles and aliphatic sulfur compounds) than propane-1,2-diol system (17). Reineccius and Liardon (18) studied the degradation products from thiamine at lower temperatures (40, 60, and 90C) at pH 5, 7 and 9, respectively. At pH 5 and 7, the meaty compounds 2-methyl-furan-3-thiol and bis(2-methyl-3-furyl) disulfide with various thiophenes were the major reaction products, while at pH 9 the meaty compounds were not significant and the thiophenes predominated. Additional thiamine degradation products reported to be meaty include 3-mercapto-2-pentanone, 2-methyl-4,5-dihydro-furanthiol, 2-methyltetrahydrofuran-3-thiol, 4,5-dimethyl thiazole and 2,5-dimethylfuran-3-thiol. To date, the degradation products 2-methyl-furan-3-thiol, bis(2-methyl-3-furyl) disulfide, and thiazole have been found in beef. The degradation product 5-hydroxy-3-mercapto-2-pentanone is key intermediate in the formation of a number of furanthiols and thiophenes.

Lipid Oxidation

Lipid oxidation/degradation of saturated and unsaturated fatty acids leads to the formation of many aliphatic hydrocarbons, alcohols, aldehydes, ketones, acids, lactones, 2-alkylfurans and esters. Other lipid-derived flavors are the benzenoids benzaldehyde, benzoic acid, alkylbenzenes and naphtalene. Lipid oxidation starts in raw beef and continues during cooking. Mottram et al. (20, 21) and later MacLeod (22) demonstrated that the intramuscular lipids (not the adipose tissues) are responsible for the formation of most of the lipid-derived volatiles.

Non-volatile precursors of process flavor aromas are formed upon hydrolysis of substrate biopolymers, and include peptides, amino acids, nucleotides, vitamins, and reducing sugars. These compounds contribute to sweet, salty, sour, bitter and umami sensations. While the sugars contribute to sweet taste, and the acid to sour taste exclusively, the amino acids and simple peptides elicit all the 5 primary taste sensations. Most of the aroma-active compounds encountered in process flavors are generated upon heating from non-volatile precursors (23) as an effect of lipid oxidation and degradation, and from degradation and interactions of sugars, amino acids, ribonucleotides, proteins, pigments and vitamins (figure 4). Over 10,000 flavor compounds have been identified and range from the simple hydrocabons, aldehydes, ketones, alcohols, carboxylic acids, esters, ethers, to the more complex lactones, furans, pyrroles, pyridines, pyrazines, thiophenes, thiazoles, oxazoles, and other sulfur and nitrogen-containing substances.

Meat Flavors

Raw meat has only a weak sweet aroma resembling serum, and a salty, metallic, bloody taste. However, it is a rich reservoir of compounds with taste properties and aroma precursors (24). Each meat has a distinct flavor characteristic. The flavors of distinct meat species or species-specific flavors are often carried by the lipid fraction (25). For example 4-methyloctanoic and 4-methylnonanoic acids are specific to mutton while (E,E)-2,4-decadienal is specific to poultry meat. 12-Methyltridecanal has been identified as species-specific odorant in stewed beef, and is responsible for the tallowy and beef-like smell (26). The distinctive pork-like or piggy flavor noticeable in lard has in part been attributed to p-cresol and isovaleric acid (27,28,29).

The major precursors in meat flavors are the water-soluble components such as carbohydrates, nucleotides, thiamine, peptides, amino acids, and the lipids, and Maillard reaction and lipid oxidation are the main reactions that convert these precursors in aroma volatiles. The thermal decomposition of amino acids and peptides, and the caramelization of sugars normally require temperatures over 150C for aroma generation. Such temperatures are higher than those normally encountered in meat cooking. During cooking of meat, thermal oxidation of lipids results in the formation of many volatile compounds. The oxidative breakdown of acyl lipids involve a free radical mechanism and the formation of

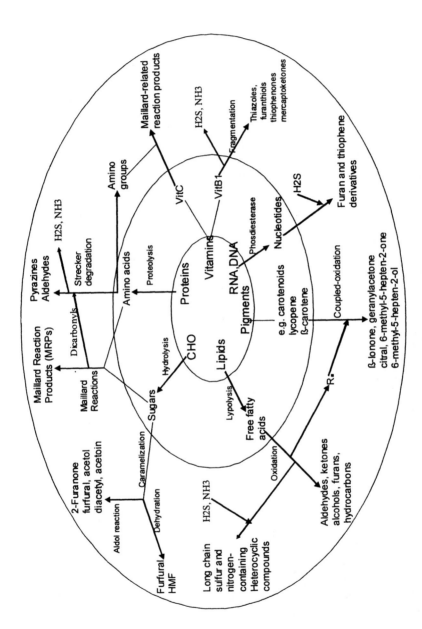

Figure 4. Wheel of Chemical Reactions Important to Flavor Formation (23)

hydroperoxides which subsequently decompose into aroma compounds. Phospholipids contain high amount of unsaturated fatty acids, and are major contributors to lipid oxidation products (13).

Poultry Flavor

In poultry meat, Maillard reactions and lipid oxidation are major sources of volatile flavor compounds (30). The meaty flavor of chicken broth has been found to be due to 2-methyl-3-furanthiol (31). This compound has also been recognized as a character impact compound in the aroma of cooked beef (30,31). However, a major difference between beef and chicken is that bis(2-methyl-3-furyl) disulfide, the oxidation product of 2-methyl-3-furanthiol and responsible for a meaty note, and methional which is responsible for a cooked potato attribute, predominate in beef whereas volatiles from the oxidation of unsaturated lipids, in particular E,E-2,4-decadienal possessing a fatty taste, and gamma-dodecalactone possessing a tallowy, fruity taste, prevail in chicken (23). Schroll and co-workers (32) found that the most abundant aldehydes in chicken flavor are hexanal and 2,4-decadienal, and that the enzymatic hydrolysis of chicken with papain increased the concentration of 2,4-decadienal from 5.2 to 13.7 mg/Kg. Hexanal and 2,4-decadienal are primary oxidation products of linoleic acid. The autoxidation of linoleate generates 9- and 13-hydroperoxides. Cleavage of 13-hydroperoxide leads to hexanal, and the breakdown of 9-hydroperoxide generates 2,4-decadienal (33). The 2,4-decadienal is a more potent odorant with threshold value of 0.00007 mg/kg compared to hexanal (0.0045 mg/kg). The major volatile compounds present in poultry meat include hexanal, 3-octen-2-one, 1-pentanol, pentanal, heptanal, octanal, 1-heptanol, 1-octanol, 1-octen-3-ol. No pyrazines were found in chicken broth, but in fried or roasted chicken, indicating that low moisture and high temperature favor the generation of pyrazines.

Pork Flavor

Lipid-derived volatile compounds dominate the flavor profile of pork cooked at temperatures below 100°C. The large numbers of heterocyclic compounds reported in the aroma volatiles of pork are associated with roasted meat rather than boiled meat where the temperature does not exceed 100°C (34,35). Of the volatiles produced by lipid oxidation, aldehydes are the most significant flavor compounds (35,36). Octanal, nonanal, and 2-undecenal are oxidation products from oleic acid, and hexanal, 2-nonenal, and 2,4-decadienal are major volatile oxidation products of linoleic acid.

Beef Flavor

It is estimated that cooked beef has 880 volatiles (37). Of these volatile compounds, only 25 have been reported to have meat odor (38,39). Aroma components are formed from non-volatile precursors on cooking as a result of lipid oxidation and Maillard reaction and degradation of thiamin, and inter-reactions of degradation products from sugars, amino acids, ribonucleotides and proteins. Meaty compounds form preferably at pH 2.2 and 5.1 rather than pH 7.1 (40). Several compounds reported to be meaty were generated at pH 2.2 and the pH of meat (pH 5.5). 4-Hydroxy-5-methyl-3(2H) furanone and 2,5-dimethyl-4-hydroxy-3(2H) furanone have been isolated from beef (but not from any other meats) and are important aroma precursors (23). Natural precursors of 4-hydroxy-5-methyl-3(2H) furanone in beef are ribose-5-phosphate and can be formed upon heating at 100C for 2.5 hours at pH 5.5 (23). Ribose-5-phosphate can be obtained from nucleotide by heating at 60C or during autolysis in muscle. Precursors of 2,5-dimethyl-4-hydroxy-3(2H) furanone are optimum at pH 4 and decomposes rapidly with temperature (41, 23). These furanone react readily with hydrogen sulfide to generate a host of heterocyclic compounds containing oxygen and/or sulfur atoms. 2-Alkyl-(2H)-thiapyran and 2-alkyl-thiophene were found in the volatiles of cooked beef and lamb trimmed fat cooked at 140C for 30 min (42). 2-Alkyl-(2H)-thiapyrans were formed when 2,4-decadienal was reacted with H2S at 140C for 30 min. Grosch and coworkers have developed a screening protocol to select beef aroma components of high sensory significance (31).

Lipid oxidation starts in raw beef and continues during cooking. Mottran et al. (21) demonstrated that the intramuscular lipids (not the adipose tissues) are responsible for the formation of most of the lipid-derived volatiles. Intramuscular lipids consist of marbling fat made primarily of triglycerols and structural or membrane lipids made of phospholipids. The phospholipids contain relatively high amount of unsaturated fatty acids more prone to oxidation. In beef, the intramuscular tissue phospholipids are sufficient in imparting a full meaty aroma (22).

Sheepmeat

Unlike the taboos that apply to beef (Hindu) and pork (Moslem, Jewish), there are no religious or cultural taboos on eating sheepmeat. In spite of this fact, consumption of sheepmeat remains low because of objectionable odor and waxy mouthfeel upon ingestion due to high melting point fat (46). The objectionable goaty and muttony odors have been attributed to the branched chain fatty acids 4-methyloctanoic, 4-ethyloctanoic and 4-methylnonanoic acids (47,24).

Hydrolysis of triglycerides or other precursors accounted for most of these acids. Ha and Lindsay (48) proposed that alkylphenols contributed to sheepmeat odor. Subsequently, Young et al., (49) reported that 4-methylphenol was correlated with animal odor. It has been reported that mutton aromas contain higher concentrations of 3,5-dimethyl-1,2,4-trithiolane and thialdine (2,4,6-trimethylperhydro-1,3,5-dithiazine) as compared to those of others species. Other sulfur-containing compounds were also found in higher amount and this observation was attributed to the high content of sulfur-containing amino acids in mutton compared with those in beef and pork (24). Kunsman and Raley (50) found that the depot fat tissues of lamb produced considerably more hydrogen sulfide than those of beef. In sheepmeat, glutathion is the sulfur precursor implicated in hydrogen sulfide formation (51).

Seafood Flavors

The flavor of fish and seafoods is composed of taste-active low molecular-weight extractives and aroma-active compounds. The taste-active compounds are more abundant in the tissues of mollusks and crustaceans than fish. The most important non-volatile taste components are free amino acids, nucleotides, inorganic salts and quaternary ammonia bases. Alcohols, aldehydes, ketones, furans, nitrogen-containing compounds, sulfur-containing compounds, hydrocarbons, esters and phenols are the most important volatiles is shellfish. Alkyl pyrazines and sulfur-containing compounds are important contributors to the cooked aroma of crustaceans. Furans pyrazines, and Lactones have been found in heat-treated seafoods. Dimethyl disulfide, dimethyl trisulfide, heterocyclic sulfur-containing compounds (alkylthiophenes) have been found in most thermally treated crustaceans like prawn, crab, oyster, crayfish and shrimp (52).

The aromas associated with very fresh fish are usually mild, delicate and fresh (53,54), and generally described as green (hexanal), melon-like ((E,Z)-3,6-nonadienal), iodine-like (bromophenols). Fresh fish and seafood aromas are due to volatile carbonyls and derive from lipoxygenase catalyzed oxidation of polyunsaturated fatty acids. The oxidation of Eicosapentaenoic acid (C20:5) leads to C5 to C9 alcohols, aldehydes, ketones and hydrocarbons. The formation of methyl mercaptan, dimethyl sulfide and dimethyl disulfide in fresh fish at the time of harvest has been reported by Shiomi et al. (55). Although these compounds are usually associated with fish deterioration, they contribute to the fresh aroma character at low concentrations. For instance, dimethyl sulfide is known to provide a pleasant seashore-like smell while dimethyl disulfide gives a pleasant crab-like aroma at concentrations less than 100 ppb. Low concentrations of bromophenols may be responsible for the desirable brine-like or sea-like aroma associated with salt-water fish (56).

Fish deterioration arise from the reduction of trimethylamine oxide (TMAO), autoxidation, through contamination from environmentally derived flavors, such as Geosmin and 2-methylisoborneol, and processing. TMAO has no odor whereas TMA is described as old fish or fish house-like. Autoxidation of polyunsaturated fatty acids eicosapentaenoic and docosahexanoic acids forms 2,4,7-decatrienal and other aldehydes. 2,4,7-Decatrienal is associated with fishy, cod liver oil-like aromas. Geosmin and 2-methylisoborneol are produced by microorganisms in water, and elicit musty or earthy flavors. Alkyl- and thiophenols from industrial waste are responsible for offensive taint odors in fish.

Hydrolyzed Vegetable Proteins

Hydrolyzed vegetable proteins (HVP) have been used to impart savory and meaty flavor to various foods. They are known to enhance and intensify naturally occurring savory flavors, and generally round off and balance the savory characteristics of the food material. Unfortunately, they lose attractiveness as food ingredient because of the high content of salt, monosodium glutamate, monochloropropanols and dichloropropanols. The propanols are known carcinogens, and this has made food processors limit the use of HVP in new product development. Furthermore, at temperatures above 200C, they become unstable and lose their typical flavor with the development of caramelized and salty taste on prolonged heating.

Wheat gluten, corn gluten, defatted soy flour, defatted peanut flour, defatted cotton seed flour are sources of proteins for HVP manufacturing. A typical HVP process calls for hydrolysis with HCL for 11 hours at 212F or 1.5 hours at 250F. The process mixture is a concentrate containing 23% water, 32% HCL (at 35%), 13% NaOH, and 32% flavor solids rich in proteinaceous material (57). Under the conditions used in hydrolysis, the amino acids from proteins react with carbohydrates through the well-known Maillard reactions to produce meaty flavors. The hydrolysate is neutralized with NaOH to pH 4.8 to 6.0. To meet customer demand, salt, MSG, caramel color, and other food additives are often added to HVP.

Yeast Extracts and Autolysates

Typical flavor notes that yeast extracts and autolysates (YEA) impart to foods are cheesy, meaty or savory. YEA finds application in such products as meats, meat dishes, soups, boullions, sauces, gravies, cheese, barbecue, salad dressings, seasoning salts, spreads, etc. Aside from flavor enhancement, yeast products are used as texturizers, stabilizers and thickeners. Yeast products are also well known for their nutritional benefits, and are incorporated into many foods as ingredients for nutritional fortification due to their relatively high content of proteins, vitamins (B1, B2 and nicotinic acid) and minerals (58). YEA are products of self-digestion or autolysis of yeast cells (baker's, torula, brewer's). During the autolytic process, yeast endogenous enzymes hydrolyze yeast proteins, nucleic acids and carbohydrates to their monomer units with taste properties. Different savory and meaty notes are developed by varying the autolysis conditions. Autolysis results in a slurry called yeast autolysate. The removal of cell walls and bitter tasting compounds from the slurry yields a liquid known as yeast extract. The organoleptically desired odors of yeast extracts are predominantly developed during heat processing by complex series of reaction involving Maillard reactions, lipid oxidation, and thermal degradation of fat, sugars, amino acids, thiamin, nucleotides, etc. Heterocyclic flavor compounds such as thiazoles, pyrroles, pyrazines, and trithiones, formed during processing, have the greatest impact on final flavor (59).

Coffee Flavors

Coffee flavors form during roasting from dicarbonyl compounds which derive from carbohydrates. The thermal degradation of hexoses is thought to be the precursors of furanones like HDMF. The presence of alkylpyrazines affords the characteristic roast notes. These pyrazines are formed through Strecker degradation and the condensation of the resulting Strecker products (60).

The astringency and the mouth feel of coffee is partly attributed to the presence of phenols which are thought to be formed by the degradation of feruloyl quininc acid (61).

Thiols are another key group of odor compounds in coffee, these are formed due to Strecker degradation of amino acids like methionine to give methional and the reaction of H2S (formed by the degradation of Cysteine) with furaldehydes, namely the formation of 2-furylthiol (62).

Coffee flavor has been subject of several studies (63, 64, 65, 66, 67, 68). Over 800 flavor compounds have been identified in coffee, and some of those important to flavor are listed in table I. There is evidence that coffee flavor is unstable. Kumazu and Masuda (66) reported that the sulfur compounds 2-furfurylthiol, methional, 3-mercapto-3-methylbutyl formate decreased in coffee brew during heating, while methanethiol, acetic acid, 3-methylbutanoic acid, 2-furfuryl methyl disulfide and 4-hydroxy-2,5-dimethyl-3(2H) furanone increased. Melanoidins in coffee brew are also involved in the loss of 2-furfurylthiol. These observations concur with those made by Hoffmann & Schieberle (64), and Hofmann et al (63). The loss of 2-furfurylthiol through a Fenton-type reaction system was also investigated by Blank et al 69), with up to 90% 2-Furfurylthiol readily decompose to difurfuryl disulfide and trisulfide upon heating at 37C for 1 hour.

Chocolate Flavors

Cocoa aroma is crucially dependent on harvesting, fermentation, drying and roasting conditions. The fresh beans have the taste and aroma of vinegar. During roasting, lipid oxidation, Maillard reaction and Strecker degradation processes lead to aroma formation. Counet et al (70) reported that via Maillard reactions, cocoa roasting converts flavor precursors formed during fermentation to 2 main classes of odorants pyrazines and aldehydes. More than 600 flavor compounds have been identified in cocoa and cocoa products (70, 71). The most important odorants of cocoa mass, determined by AEDA, have been reported by Belitz and Grosch (72). Amino acids released from fermentation served as precursors to 3-methylbutanal (malty), phenylacetaldehyde (honey-like), 2-methyl-3(methyldithio)furan (cooked meat-like), 2-Ethyl-3,5-dimethylpyrazine (earthy, roasted) and 2,3-Diethyl-5-methylpyrazine (earthy roasted) during roasting. These authors (72) also reported the occurrence of ethyl 2-methylbutanoate (fruity), hexanal (green), 2-methoxy-3-isopropylpyrazine (peasy, earthy), E-2-octenal (tallowy), E-2-nonenal (fatty, waxy), Z-4-heptanal (biscuit-like), delta-octalactone and delta-decalactone (sweet coconut like) among the odorants of cocoa mass. Counet and co-workers (70) revealed the presence of 33 potent odorants in the neutral/basic fraction of dark chocolate by AEDA. Three of them, 2-methylpropanal, 2-methylbutanal, and 3-methylbutanal had strong chocolate odor. 2,3-Dimethylpyrazine, trimethylpyrazine, tetramethylpyrazine, dimethylethylpyrazine, diethylmethylpyrazine, and furfurylpyrrole were characterized as possessing cocoa, nutty, coffee notes.

Bread Flavors

Although more than 280 compounds have been identified in the volatile fraction of wheat bread, only a small number is responsible for the flavor notes in the crust and the crumb. Schieberle and Grosch (73) used aroma extract dilution analysis (AEDA) to select 32 odorants in wheat. Among the odorants, 2-acetyl-pyrroline (roasty, bread crust-like) was the most potent aroma, followed by E-2-nonenal (green, tallowy), 3-methylbutanal (malty, nutty), diacetyl (buttery) and Z-2-nonenal (green, fatty).
Bread flavors compounds 2-acetylpyrroline and 6-acetyl-1,2,3,4-tetrahydropyridine (cracker-like) derive from Maillard reactions. Maillard reaction products maltol and isomaltol contribute to the freshness of baked bread. Kirchoff and Schieberle 74) reported that the typical flavor of bread is formed as a result of enzymatic reactions occurring during dough fermentation by yeasts and/or lactic acid bacteria followed by thermal reactions during baking. The authors also found that 3-methylbutanal, E-2-nonenal, acetic acid, 2,4-decadienal, hexanal, phenylacetaldehyde, methional, vanillin, 2,3-butadione, 3-OH_4,5-dimethyl-2(5H) furanone and 2,3-methylbutanoic acid were important contributors to the flavor of sourdough bread crumb.

Flavor Stability

One challenge facing the flavor industry today deals with the stability of process/reaction flavors to heat, UV-light, oxidation, binding to food matrices, diffusion and interactions with the environment in which they find application. Seeventer et al. (75) studied the stability of thiols formed from model system ribose/cystein, and reported that 2-methyl-3-furanthiol, 2-furfurylthiol, 2-mercapto-3-butanone and furaneol decreased during storage. In brewed coffee,

Table I Most important chemicals responsible for the coffee flavor notes.

Compounds	Descriptors	Ethiopian Arabica (68) (CHARM-Analysis)	Columbian Arabica (68) (AEDA)	Indonesian Robusta (72) (Aroma Value)
4-Ethenyl-2-methoxyphenol (4-vinylguaiacol)	phenolic	490	256	8900
4-Hydroxy-2,5-dimethyl-3(2H)-furanone (furaneol)	sweet-caramel	410	256	5700
3-Methyl-2-buten-1-thiol	smoke-roast	360	16	28000
Beta-Damascenone	sweet-fruity	340		270000
2,3-butanedione (diacetyl)	buttery	340	32	3200
2-Furfurythiol	smoke-roast	320	6	170000
Methional	soy sauce	310	64	
3-Methylbutyric acid	acidic	280		
2- and 3-Methylbutanals	buttery	240		
2-Methyl-3-furanthiol	nutty-roast	230	16	
2-Ethyl-3,5-dimethylpyrazine	nutty-roast	230	64	5900
2-Methoxyphenol (guaiacol)	phenolic	210	128	11000
2-Methoxy-3-(2-methylpropyl)pyrazine	green-earthy	200		
3-Mercapto-3-methylbutyl formate	green-black currant	190	64	33000
2,3-Pentanedione	buttery	180	16	660
6,7-Ddihydro-5-methyl-5H-cyclopentapyrazine	nutty-roast	170		
2-Ethyl-4-hydroxy-5-methyl-3(2H)-furanone (homofuraneol)	sweet-caramel	170	128	12000

Compound	Odor			
2,3-Diehtyl-5-methylpyrazine	nutty-roast	120	64	3400
E-2-Nonenal	buttery	39		
Unknown	sweet-caramel	30		
Unknown	sulfurous, burnt		16	
Unknown	earthy		64	
4-Ethyl-guaiacol			256	360
Vanillin			32	640
3-Hydroxy-4,5-dimethyl-2(5H)-furanone (sotolon)	caramelic, curry,		64	32
5-Ethyl-3-hydroxy-4-methyl-2(5H)-furanone (abhexon)	seasoning-like			11
Isobutyl-2-methoxy-pyrazine			32	3000

decrease in amount of 2-furfurylthiol, methional, and 3-mercapto-3-methylbutyl formate was observed after heating, and has been attributed to interactions with melanoidins (63,64,66). The loss of 2-furfurylthiol was linked to Fenton-type oxidation upon heating at 37^0C for 1 hour (69).

Summary

Considerable progress has been made in characterization of process flavors, study of flavor precursors, and flavor regulation. Nevertheless, the growing demand for authentic and ethnic foods in the western world calls for renewed efforts in process and reaction flavors. Moreover, although work is underway to elucidate structure and properties of taste-active compounds in process and reaction flavors, more work remain to be done. This symposium reports recent significant findings in process and reaction flavors.

References

1. *Top 10 Flavor & Fragrance Industry Leaders Estimated Sales Volume*, Leffingwell & Associates, 2003, URL: www.leffingwell. Com.
2. *World Demand for Flavor & Fragrances*, The Freedonia Group, Cleveland, OH, 2003, URL: www.freedoniagroup.com.
3. Giese, J. *Food Technology* **2004**, 58(2):30- 33.
4. Wittewen, J. *Personal Communication*, **2003**.
5. Reineccius, G. A. In *Flavor Chemistry: Industrial and Academic Research*; Risch, S. J. and Ho, C. T., Eds; ACS Symposium Series 756; American Chemical Society: Washington, DC, 2000, pp. 13-21.
6. Manley, C.H. In *Flavor Chemistry: Industrial and Academic Research;* Risch, S. J. and Ho, C. T., Eds; ACS Symposium Series 756; American Chemical Society: Washington, DC, 2000, pp. 2-12.
7. FEMA Membership Directory, Flavor and Extract Manufacturers Association, 2003, Washington, DC.
8. Bailey, M. E. In *Flavor of Meat, Meat Products and Seafoods*; Shahidi, F., ed.; Blackie Academic & Professional: New York, NY, 1998, pp. 267-289.
9. Mottram, D. S. In *Thermally Generated Flavors: Maillard, Microwave and Extrusion Processes*; Parliament, T. H., Morello, M. J. and McGorrin, R. J., eds.; ACS Symposium Series 543; American Chemical Society: Washington, DC, 1994, pp. 104-141.
10. Nursten H. E. In *Developments in Food Flavors*, Birch G.G. and Lindley, M. G., Eds; Elsevier, London, 1986, pp173-190.

11. Hurrell, R. F. In *Food Flavors;* Morton, I. D. and MacLeod, A. J., Eds; Elsevier, Amsterdam, 1982, pp. 399-437.
12. Mottram, D. S. In *Flavor of Meat, Meat Products and Seafoods*; Shahidi, F., ed.; Blackie Academic & Professional: New York, NY, 1998, pp. 5-26
13. Schutte, L. *CRC Critc Rev. Food Technol.* **1974**, 4, 457-505.
14 .Buchholz, L. L., Jr. In *Thermal Generation of Aromas*; Parliament, T. H., McGorrin, R. J. and Ho, C. T., Eds; ACS Symposium Series 409; American Chemical Society: Washington, DC, 1989, pp. 406-420.
15. Ho, C.-T. In The Maillard Reaction: Consequences for the Chemical and Life Sciences; Raphael Ikan, ed.; John Wiley & Sons, Ltd, 1996.
16. Dwievedi, B. K. and Arnold R. G. *J. Agric. Food Chem.* **1973**, 21, 54-60.
17. Hartman, G. J. Carlin, J. T., Scheide, J. D. and Ho, C. T. *J. Agric. Food Chem.* **1984**, 32, 1015-1018.
18. Reineccius, G. A. and Liardon, R. In *Topics in Flavor Research*; Berger, R. G. Nitz, S. and Schreier, P., Eds; Eichhorn, Marzling-Hangenhan 1985, pp. 125-136.
19. Guntert, M., Bruning, J. Emberger, R., Hopp, R., Kopsel M., Surburg, H. and Werkhoff, P. In *Flavor Precursor, Thermal and Enzymatic Conversions*. Teranishi, R. Takeoka, G. R. and Guntert, M., Eds; ACS Symposium Series 490; American Chemical Society, 1992.
20. Mottram, D. S. Edwards, R. A. and MacFie. H. J. H. *J. Sci. Food Agric.* **1982**, 33, 934-944.
21. Mottram, D. S. and Edwards, R. A. *J. Sci. Food Agric.* **1983**; 34, 517-522.
22. MacLeod, G. In *Flavor of Meat, Meat Products and Seafoods*; Shahidi, F., ed.; Blackie Academic & Professional: New York, NY, 1998, pp. 27-60.
23. Sucan, M. K., Byerly, E. A., Grun, I U, Fernando, L. N. and trivedi, N. B. In *Bioactive Compounds in Foods: Effects of Processing and Storage*; Lee, T.C. and Ho, C.T., Eds; ACS Symposium Series 816; American Chemical Society: Washington, DC, 2002, pp187-205.
24. Shahidi F. In *Flavor of Meat and Meat Products*; Shahidi, F., ed.; Blackie Academic & Professional: New York, NY, 1998.
25. Mottram, D. S.; Edwards, R. A; McFie, H. J. *J. Sci. Food Agric.* **1982**, *33*, 934-944.
26. Guth, H; Grosch, W. *Lebensmittel-Wissenschaft-und-Technologie* **1993**, *26*, 171-177.
27. Ha, J.K.; Lindsay, R.C. *J. Food Sci.* **1991**, *56*, 1197-1202.
28. Ha, J.K.; Lindsay, R.C. *Lebensmittel-Wissenschaft und –Technologie* **1990**, *23*, 433-440.
29. Lindsay, R.C. In *Food Chemistry*; O. R. Fennema, ed.; Marcel Dekker: New York, 1996; pp 723-765.
30. Shi, H.; Ho, C.-T. In *Flavor of Meat and Meat Products*; F. Shahidi, ed.; Blackie Academic & Professional: New York, NY, 1994.

31. Gasser, U.; Grosch, W. *Z Lebensm Unters Forsch* **1990**, *190*, 3-8.
32. Schroll, W.; Siegfried, N.; Drawert, F. *Z Lebensm Unters Forsch* **1988**, *187*, 558-560.
33. Ho, C. T. and Carlin, J. T. In *Flavor Chemistry, Trends and developments*; Teranishi, R. Buttery, R. G. and Shahidi, F., Eds; American chemical Society: Washington, DC 1989, pp. 92-104.
34. Mottram, D. S. *J. Sci. Food Agric.* **1985**, *36*, 377-382.
35. Ho, C.-T.; Oh, Y.-C.; Bae-Lee, M. In *Flavor of Meat and Meat Products*; F. Shahidi, ed; Blackie Academic & Professional: New York, NY, 1994.
36. Frankel, E. N. *Prog. Lipid Res.* **1982**, *22*, 1-33.
37. Maarse, H. and Visscher, C. A. *Volatile Compounds in Foods, Qualitative and Quantitative Data*; TNO-CIVO, Zeist, The Netherlands, 1989.
38. Werkhoff, P., Emberger R., Guntert, M., Kopsel, M., Kuhn, M and Surburg, H. J. In *Thermal Generation of Aromas*; Parliament, T. H., McGorrin, R. J. and Ho, C.-T., Eds; American Chemical Society: Washington, DC 1989, pp 460-478.
39. Werkhoff, P., Bruining, P., Emberger R., Guntert, M.and Kopsel, M., Kuhn, W. and Surburg, H. *J. Agric. Food Chem.* **1990**, 38, 777-791.
40. Shu, C. K., Hagedorn, M. L. and Ho, C. T. *J. Agric. Food Chem.* **1986**, 34, 344-346.
41. Hirvi, T. Honkanen, E. and Pysalo, T. Lebensm. *Wiss. Technol.* **1980**, 13, 324-325.
42. Elmore, J. S. and Mottram, D. S. *J Agric. Food Chem.* **2000**, 48, 2420-2424.
43. Gasser, U.; Grosch, W. *Z Lebensm Unters Forsch* **1990**, *190*, 3-8.
44. Kerler, J.; Grosch, W. *Z Lebensm Unters Forsch* **1997**, *205*, 232-238.
45. Gasser, U.; Grosch, W. *Z Lebensm Unters Forsch* **1988**, *186*, 489-494.
46. Young, O. A. and Braggins, T. J. In *Flavor of Meat, Meat Products and Seafoods*; Shahidi, F., ed.; Blackie Academic & Professional: New York, NY, 1998, pp.101-130.
47. Karl, V., Gutser, J., Dietrich, A., Maas, B. and Mosanld, A. *Chirality* **1994**, 6, 427-434.
48. Ha, J.K.; Lindsay, R.C. *J. Food Sci.* **1991**, *56*, 1197-1202.
49. Young, O. A. Berdague, J.-L., Viallon, C. Rousset-Akrim, S. and Theriez, M. *Meat Sci.* **1997**, 45, 169-181.
50. Kunsman, J. E. and Raley, M.L. *J. Food Sci.* **1975**, 40, 506-508.
51. Cramer, D. A. *Food technol.* **1983**, 37(5), 249-257.
52. Spurvey, S., Pan, B. S. and Sahidi, F. In *Flavor of Meat, Meat Products and Seafoods*; Shahidi, F., ed.; Blackie Academic & Professional: New York, NY, 1998, pp. 159-196.
53. Lindsay, R. C. *Food Rev. Int.* **1990**, 6, 437-455.

54. Durnford, E. and Shahidi, F. In *Flavor of Meat, Meat Products and Seafoods*; Shahidi, F., ed.; Blackie Academic & Professional: New York, NY, 1998, pp. 131-158.
55. Shiomi, K., Noguchi, A. Yamanaka, H., Kikuchi, T. and Iida, H. Comp Biochem. Physiol. 1982, 71B, 29-31.
56. Boyle, J. L., Lindsay, R. C. and Stuiber, D. A. *J. Food Sci.* 1992, 57, 918-922.
57. Nagodawithana, T. W. In Bioprocess Production of Flavor, Fragrance, and Color Ingredients. Gabelman, A., ed; John Wiley & Sons, Inc., New York 1994, pp. 135-168.
58. Dziezak,J. D. *Food Technology* February 1987.
59. Nagodawithana, T. W. *Food Technology* November 1992, pp 140.
60. Amrani-Hemaimi, M., Cerny, C., and Fay, L.B. *J Agric. Food. Chem.* 1995, 43. 2818.
61. Tressl, R. In *Thermal Generation of Flavors*; Parliament,T. H. McGorrin, R. J. and Ho, C.T., Eds; ACS Symposium series 409, American Chemical Society: Washington, DC. 1989, pp.285-301.
62. Grosch, W. In *Proceedings of the 18ᵗʰ ASIC Colloquium*; Helsinki., ASIC, Paris. , France, 1999.
63. Hofmann, T., Czerny, M., Calligaris, S. and Schieberle, P. *J. Agric. Food Chem.* 2001, 49, 2382-2386.
64. Hofmann, T. and Schieberle, P. J. *Agric. Food Chem.* 2002, 50, 319-326.
65. Czerny, M. and Grosch, W. *J. Agric. Food Chem.* 1999, 47, 695-699.
66. Kumazawa, K. and Masuda, H. *J. Food Chem.* 2003, 51, 2674-2678.
67. Stadler, R. H., Varga, N., Milo, C., Schilter, B., Vera, F. A. and Welti, D. H. *J. Agric. Food Chem.* 2002, 50, 1200-1206.
68. Akiyama, M., Murakami, K, Ohtani, N., Iwatsuki, K., Sotoyama, K. Wada, A., Tokuni, K., Iwabuchi, H. and Tanaka, K. *J. Agric. Food chem.* 2003, 51, 1961-1969.
69. Blank, I., Pascual, E. C., Devaud, S., Fay, L. B., Stadler, R. H., Yeretzian, C. and Goodman, B. *J. Agric. Food Chem.* 2002, 50, 2356-2364.
70. Counet, C., Callemien, D. Ouwerx, C. Collin, S. *J. Agric. Food Chem.* 2002, 50, 2385-2391.
71 Schnermann, P. and Schieberle, P. *J. Agric. Food Chem.* 1997, 45, 867-872.
72. Belitz, H.-D. and Grosch, W. Food Chemistry , Springer-Verlag, Berlin 1999, pp. 875.
73. Schieberle, P. and Grosch, W. In *Thermal Generation of Flavors*; Parliament,T. H. McGorrin, R. J. and Ho, C.T., Eds; ACS Symposium series 409, American Chemical Society: Washington, DC. 1989, pp.258-267.
74. Kirchoff, E. and Schieberle, P. *J. Agric. Food Chem.* 2001, 49, 4304-4311.
75. Seeventer, P. B. Weenen, H., Winkel, C. and Kerler, J. *J. Agric. Food Chem.* 2001, 49, 4292-4295.

Regulatory and Safety Aspects of Process Flavors

Chapter 2

The Safety Assessment of Process Flavors

John B. Hallagan

The Flavor and Extract Manufacturers Association of the United States, 1620 I Street, N.W., Suite 925, Washington, DC 20006 (202–331–2333)

Introduction

Process flavors are widely used in prepared foods to enhance the taste and odor of these products. Process flavors were identified as potential dietary sources of heterocyclic amines, both monocyclic heteroaromatic amines (MHAs), and polycyclic heteroaromatic amines (PHAs). MHAs are innocuous substances, some of which have flavoring characteristics. However, PHAs are reported to be potent mutagens and animal carcinogens (Munro *et al.*, 1993) and have been the subject of significant interest regarding their potential human health risk.

An assessment of the potential presence of PHAs in process flavors, and the implications for the safety assessment of these flavors, was sponsored by the Flavor and Extract Manufacturers Association of the United States (FEMA). The assessment was conducted in three stages: (1) identification and quantification of PHAs in process flavors; (2) identification of the food categories which could potentially contain process flavors and calculation of their respective daily *per capita* intakes; and (3) determination of the daily *per capita* intake of PHAs through consumption of foods identified as sources of naturally-occurring PHAs compared to foods containing process flavors.

This report describes work done during the 1990s to provide an indication of the degree to which PHAs are present in process flavors through normal manufacturing processes. Much of the analytical work was performed by the staff at the Lawrence Livermore National Laboratory - Mark Knize and Cindy Salmon were the primary investigators on this project and deserve much of the credit for the well-designed and well-conducted studies reported here.

Process Flavors

Process flavors are most often prepared by the heat processing of foods or food ingredients followed by isolation and purification. In general, process flavors mimic flavors present naturally in cooked foods. Process flavors are used at low levels because of their powerful organoleptic qualities. Manley (1999), and Janiec and Manley (2003) have reviewed various aspects of process flavor production and use.

Process flavors are generated from interactions between protein nitrogen (amino acids/peptides), carbohydrate, and fat or fatty acid sources during thermal processing. These thermally driven chemical reactions can generate desirable flavoring compounds which may include MHAs. However, under certain conditions of temperature and cooking time the potential exists to form PHAs. MHAs often formed during the cooking of foodstuffs include pyrazines, pyridines, and thiazoles. PHAs formed during cooking include imidazoquinolines, imidazoquinoxalines, and imidazopyridines. Skog et al. (1998) reviewed the available literature and provided a detailed review of the various parameters that may be involved in the formation of PHAs in cooked foods.

Polycyclic Heteroaromatic Amines (PHAs)

Intermediate compounds formed during cooking include dihydropyridines and dihydropyrazines. Dihydropyridines and dihydropyrazines constitute precursors from which MHAs or PHAs can be formed. In the presence of creatinine (an amino acid found endogenously in all animals), dihydro compounds can react to form PHAs. Because creatinine occurs almost exclusively in meat and fish products, PHAs are predominantly formed during the heat processing of animal products. It has been reported that the formation of PHAs is associated with the type of meat product, cooking method and degree of doneness/surface browning, and that meats, which are cooked over a longer duration and to a high degree of surface browness, contain higher levels of PHAs (Sinha *et al.*, 1998a,b).

PHAs identified in cooked foods that are of concern to human health include 2-amino-3-methylimidazo[4,5-*f*]quinoline (IQ), 2-amino-3,4-dimethyl-3*H*-imidazo[4,5-*f*]quinoline (MeIQ), 2-amino-3,8-dimethylimidazo[4,5-*f*]quinoxaline (MeIQx), 2-amino-2,4,8 (or 3,7,8)-trimethylimidazo[4,5-*f*]quinoxaline (diMeIQx), 2-amino-*N*-methyl-5-phenylimidazopyridine (PhIP), 3-amino-1,4-dimethyl-5*H*-pyrido[4,3-*b*]indole (Trp-P-1), 3-amino-1-methyl-5*H*-

28

pyrido[4,3-*b*]indole (Trp-P-2), 2-amino-6-methyldipyrido[1,2-*a*:3'2'-*d*]imidazole (Glu-P-1), 2-aminodipyrido[1,2-*a*:3'2'-*d*]imidazole (Glu-P-2), 2-amino-9*H*-pyrido[2,3-*b*]indole(AaC), and 2-amino-3-methyl-9]indole (A*H*-pyrido[2,3-*b*C).]indole (MeAaC). An extensive review of the toxicological data on these PHAs indicates that these compounds possess carcinogenic potential of varying potencies (Munro *et al.*, 1993). Keating and Bogen (2001) suggest that five PHAs - AaC, IQ, DiMeIQx, MeIQx, and PhIP, are predominant in terms of human exposure.

Munro *et al.*, (1993) reviewed metabolic and pharmacokinetic studies in laboratory animals which showed that PHAs are rapidly absorbed, and that once absorbed these substances or their metabolites are distributed to various organs such as the liver, kidneys, intestine, stomach, and lungs. PHAs undergo metabolic activation or detoxication and are eliminated through either the urine or feces to more or less the same degree. The amounts of PHAs or their metabolites in tissues have been reported to decline to undetectable levels within 72 hours; however, residual levels have been detected in both liver and intestines (Bergman, 1985; Alldrick and Rowland, 1988; Gooderham *et al.*, 1991).

PHAs are highly reactive in biological systems and have been reported to possess biochemical characteristics consistent with electrophilic carcinogens. A review of genotoxicity data indicate that PHAs demonstrate mutagenic activity in bacterial, insect and mammalian *in vitro* and *in vivo* systems (Munro *et al.*, 1993). Carcinogenicity studies on eight of the nine PHAs reviewed by Munro *et al.*, (1993) showed that PHAs are multiple site carcinogens in rodents with only one, PHA (Trp-P-1), reported to be solely a hepatic carcinogen. However, these carcinogenicity studies are of little value in estimating the risk to consumers from low exposures such as those encountered through food consumption because they were performed in laboratory animals fed high dietary levels of PHAs and therefore do not provide data on low-dose exposures adequate to characterize dose-response relationships for PHAs.

PHAs in Food
PHAs occur naturally at low levels in cooked foods and are typically found in parts per billion (ppb) levels. Assessment of the concentrations of PHAs was once limited due to constraints in the analytical methodology used to quantify PHAs at these low levels. A solid phase extraction and a high pressure liquid chromatography (HPLC) method coupled with ultraviolet detection was developed by Gross (1990) as an early, accurate method. Others have since refined the methodology further, including Knize and Salmon (1998), who used their method to generate the data described in this report, and Solyakov et al. (1999) and Richling et al. (1999).

Food categories containing naturally occurring PHAs were identified and maximum and minimum levels of PHAs in these foods were determined from the literature. Using these data and the three-day average *per capita* intake of various types of cooked meats the total *per capita* maximum, minimum and median intakes of PHAs from consumption of cooked meats were determined and are summarized in Table 1. The data indicate that the concentrations of PHAs in cooked meats are variable and may be associated with limitations in the studies used to report these concentrations. Some studies have based their levels of PHAs on a single analysis of the food sample while other studies did not detect or quantify all of the PHAs in each food sample assayed. There are little or no data for some PHAs in certain food categories, which creates the potential for underestimation of the significance of these PHAs in contributing to naturally-occurring PHAs in the diet.

Dietary Exposure to Process Flavors
Intake of process flavors from foods in various categories was determined based on the Nationwide Food Consumption Survey (NFCS) database, which was derived from surveys of U.S. food consumption conducted by the U.S. Department of Agriculture (USDA) in 1977-1978. Although 1987-1988 data were available, the 1977-1978 data were considered to be more representative of the general population (n= 30,770; n = 28,006 for individuals providing a 3-day diet history) than the more recent 1987-1988 data (n = 10,193 total; n = 8,224 with a three-day diet history). Data were not available for the years after 1988 at the time the analyses reported here were performed.

USDA surveys are designed as self-weighting, multistage, stratified area samples of the U.S. population, meaning that the USDA, in conducting this survey, determined the relative proportions of different demographic groups in the entire population, and invited participation of each demographic group in the survey in those proportions. If all individuals had responded, the survey results would be directly representative of the U.S. population. However, there was non-response to the survey, at different rates in the different demographic groups, requiring that the data derived from the NFCS are weighted in order for it to be considered representative of the U.S. population. This technique of weighting to correct for non-response has been established by the USDA where groups with low response rates are given greater weight, so their representation in the final survey of sample is proportional to their representation in the general population.
When the categories of food are specific, such as those used in this assessment, the number of individual respondents in each category is often very low. With few respondents, there is an increased chance that the responses obtained do not accurately represent the average of the general population or demographic

Table 1   Total *Per Capita* Intakes of HCAs from Cooked Meats (Based on 3-Day Average Food Intakes)

Compound	Intake (ng/kg bw/day)		
	Minimum	Maximum	Median[a]
IQ	0.36	150	1.9
MeIQ	1.3	71	5.5
MeIQx	8.9	33	16
DiMeIQx	5.2	11	6.0
Trp-P-1	0.18	56	40
Trp-P-2	0.068	13	0.24
Glu-P-1	0	0	0
Glu-P-2	240	240	240
PhIP	65	220	57
AαC	2.1	820	290
MeAαC	54	66	60

[a]Defined as the middle value (or midpoint between two middle values) on a ranked list of concentrations.

group, instead, the responses might represent those at the extremes of the population or demographic group. Therefore, in this assessment, because of the low number of individuals in most food categories, weighting was not used, in order to avoid the over-emphasis of extreme eating behaviors.

Using the NFCS database, the mean daily intakes (in g/day) of the food categories containing process flavors by persons for whom there were records for three days of food intake were determined. Because process flavors are added during the processing of foods, the categories in which they are expected to be used are complete meals or courses of a meal, as opposed to the raw materials or unprocessed foods. The categories selected for inclusion in this assessment fell under the general headings of:

Frankfurters, sausages, lunchmeats and meat spreads
Meat, poultry or fish (with sauce)
Meat, poultry or fish with starch item and sauce
Meat, poultry or fish with starch item and vegetables
Meat, poultry or fish with vegetables
Frozen plate meals with meat, poultry or fish as the major ingredient
Soups, broths or extracts with meat, poultry or fish base
Gravy from meat, poultry or fish base
Meat substitutes, mainly vegetable protein
Dried peas, lentils and mixtures
Grain products and mixtures
Pastas
Salty snacks from grain products
Frozen plate meals with grain products as major ingredient
Soups with grain products as main ingredient; tomato sauces
Vegetable soups
Salad dressings

There was a total of 192 food categories of the NFCS selected for inclusion in the assessment of the consumption of foods containing process flavors. Of these, eighteen could not be assessed because the categories did not exist in the 1977-1978 database. A subset of "convenience food" preparations (e.g. frozen dinners, macaroni and cheese dinners) also was assessed from the 1987-1988 database, in order to make an estimation of the changes in food consumption of these categories over the interval 1977-78 to 1987-88. Comparison of the consumption of convenience meals between 1977-1978 and 1987-1988 indicates that the levels of consumption of these food items has not changed significantly. There was a total of 192 food categories of the NFCS selected for inclusion in the assessment of the consumption of foods containing process flavors. Of these, eighteen could not be assessed because the categories did not exist in the 1977-1978 database. A subset of "convenience food" preparations (e.g. frozen dinners, macaroni and cheese dinners) also was assessed from the 1987-1988 database, in order to make an estimation of the changes in food consumption of these categories over the interval 1977-78 to 1987-88. Comparison of the consumption of convenience meals between 1977-1978 and 1987-1988 indicates that the levels of consumption of these food items has not changed significantly.

Analysis of Process Flavors for the Presence of Polycyclic Heteroaromatic Amines

Samples of process flavors were analyzed for the presence of five PHAs: IQ, MeIQ, MeIQx, DiMeIQx, and PhIP. The samples were analyzed using the method developed by Gross (1990) and modified by Knize and Salmon (1998). The analysis of the samples complied with the FDA's Good Laboratory Practice regulations (21 CFR Part 58).

A total of 102 process flavors were analyzed representing the variety of process flavors commonly added to foods in the United States. The study included an analysis of PHA levels in "spiked" and "unspiked" samples of process flavors to verify the accuracy and precision of the analytical method. Results of this proficiency test indicate that the likelihood of false positives or false negatives was low, and that measurement of the PHA levels would be precise and accurate (Knize and Salmon, 1998).

Of the 102 process flavor samples analyzed, 90 samples had undetectable levels of PHAs. Two samples contained one PHA each at levels barely above the 50 ppb limit of detection established for this method; one sample contained IQ at 67 ppb and the other contained MeIQx at 59 ppb. Ten other samples were found to contain one or two PHAs per sample in trace amounts below the 50 ppb limit of detection. Overall, it was reported that this assay demonstrated good sensitivity and reproducibility for detection of PHAs in process flavors and a limit of detection was established to be 50 ppb.

Using the two samples with PHA levels >50 ppb as "worst case scenarios" for exposure to PHAs through process flavor consumption, their estimated daily *per capita* intakes were calculated.

Potential Exposure to PHAs from Process Flavors

Using information on daily *per capita* intake of foods containing process flavors combined with analytical data on the levels of PHAs present in process flavors, the estimated *per capita* daily intakes of PHAs were calculated and are shown in Table 2. Since IQ and MeIQx were identified in two samples and were present at levels above the detection limit of 50 ppb, they were used as representatives of the maximum levels of exposure to PHAs from process flavors.

In Table 2, the food categories and *per capita* food consumption of relevant foods expected to contain IQ and MeIQx (g/day) are reported. The potential level of use of the process flavor in the final food product (0.5%) and the potential level of use of IQ (Sample 1) and MeIQx (Sample 2) in the PF food product (small use=25%; medium use=50%; large use=75%) are represented in Columns 3 and 4, respectively. Column 5 represents the poundage ratio which assumes that there would be approximately 6000 lbs of samples 1 and 2 in the 13 million lbs of total PF used by consumers per year. The intakes of IQ (sample 1) and MeIQx (sample 2) (ng/day) are reported in Columns 6 and 7, respectively, and are calculated based upon the usage levels reported in columns 2, 3, 4 and 5 with the assumption that IQ (Sample 1) is present at a level of 67 ng/g and MeIQx is present at a level of 59 ng/g in the food categories containing process flavors.

In summary, the total daily *per capita* intakes for IQ and MeIQx from all food categories containing process flavors were estimated to be 1.62×10^{-3} and 1.43×10^{-3} ng/day, respectively.

Safety Assessment of Process Flavors
A review of toxicology data on nine PHAs considered to be representative of PHAs present in prepared foods was conducted by Munro *et al.* (1993) and indicates that PHAs demonstrate genotoxic and carcinogenic potential. It appears that PHAs require metabolic activation in order to exert their genotoxic potential. Studies with *Salmonella typhimurium* TA98 show differences in mutagenic potential among PHAs (de Meester, 1989; Grivas *et al.*, 1985; Shioya *et al.*, 1987). Overall, it appears that *in vitro* genotoxicity data have not been adequate predictors of genotoxic potency of PHAs in whole animals because metabolic processes and pharmacokinetic factors have an impact on the potency of PHAs *in vivo*.

Carcinogenicity studies show that PHAs have the potential to induce various forms of cancer in rodents after oral administration (Matsukura *et al.*, 1981; Ohgaki *et al.*1984a,b; Takayama *et al.*, 1984; Ohgaki *et al.*, 1985; Sugimura, 1985; Takayama *et al.*, 1985; Ohgaki *et al.*, 1987; Kato *et al.*, 1988; Kato *et al.*1989; Ito *et al.*, 1991; Adamson et al., 1994). Unfortunately these studies are not useful in estimating the risk in humans of exposure to PHAs because none of the studies provide data adequate to characterize dose-response relationships. Furthermore, the high doses of PHAs administered to animals are not representative of the low levels of PHAs that occur naturally in human foods.

Table 2: Daily Per Capita Intake of IQ & MeIQ

PER CAPITA FOOD CONSUMPTION DATA (1977)

Foods Expected to Contain Processed Flavors

	(1) Food Description	(2) Food Per capita (g/day)	(3) PF Use Level (%)	(4) Potential Use in Food (%)	(5) Poundage Ratio#	(6) Sample 1 (IQ) (ng/day)	(7) Sample 2 (MeIQx) (ng/day)
1	beef with gravy	1.1354	0.5	25	4.62E-04	4.39E-05	3.86E-05
2	beef with (mushroom) soup (mixture)	0.5397	0.5	25	4.62E-04	2.09E-05	1.84E-05
3	beef with other sauces (curry, BBQ, sweet and sour, tartar)	0.3585	0.5	25	4.62E-04	1.39E-05	1.22E-05
4	pork with gravy	0.0541	0.5	25	4.62E-04	2.09E-06	1.84E-06
5	pork with other sauces (BBQ, sweet and sour, honey garlic)	0.152	0.5	25	4.62E-04	5.88E-06	5.17E-06
6	chicken or turkey with gravy	0.0201	0.5	25	4.62E-04	7.77E-07	6.84E-07
7	beef and noodles with gravy or sauce	11.9664	0.5	25	4.62E-04	4.63E-04	4.07E-04
8	beef and rice with gravy or sauce	0.7018	0.5	25	4.62E-04	2.71E-05	2.39E-05
9	ham and starch items with gravy or sauce	0.1183	0.5	25	4.62E-04	4.57E-06	4.03E-06
10	beef and vegetables with (mushroom) soup (mixture)	0	0.5	25	4.62E-04	0.00E+00	0.00E+00
11	beef, noodles and vegetables with (mushroom) soup (mixture)	0	0.5	25	4.62E-04	0.00E+00	0.00E+00
12	beef, noodles and vegetables with gravy	*	0.5	25	4.62E-04	0.00E+00	0.00E+00
13	beef, rice and vegetables with (mushroom) soup (mixture)	0	0.5	25	4.62E-04	0.00E+00	0.00E+00
14	beef, rice and vegetables with gravy	*	0.5	25	4.62E-04	0.00E+00	0.00E+00
15	chicken or turkey and noodles with sauce or gravy	0.4292	0.5	25	4.62E-04	1.66E-05	1.46E-05
16	chicken or turkey stew with potatoes vegetables and gravy	0.2582	0.5	25	4.62E-04	9.98E-06	8.79E-06
17	chicken or turkey, noodles, vegetables and gravy	0	0.5	25	4.62E-04	0.00E+00	0.00E+00
18	chicken or turkey, rice, vegetables and gravy	0	0.5	25	4.62E-04	0.00E+00	0.00E+00
19	chicken or turkey, rice, vegetables and (mushroom) soup (mixture)	0	0.5	25	4.62E-04	0.00E+00	0.00E+00
20	salisbury steak dinner with gravy	0.2684	0.5	25	4.62E-04	1.04E-05	9.14E-06
21	beef, sliced with gravy	0.0796	0.5	25	4.62E-04	3.08E-06	2.71E-06
22	meatballs, Swedish, with sauce	0	0.5	25	4.62E-04	0.00E+00	0.00E+00
23	pork, sliced with gravy	0	0.5	25	4.62E-04	0.00E+00	0.00E+00

24	turkey dinner, with gravy	0.4801	0.5	25	4.62E-04	1.86E-05	1.63E-05
25	Beef soups (include broth and consommé)	2.2328	0.5	25	4.62E-04	8.63E-05	7.60E-05
26	chicken soups (include broth and consommé)	2.5109	0.5	25	4.62E-04	9.71E-05	8.55E-05
27	chicken or turkey soups, cream of	0.6518	0.5	25	4.62E-04	2.52E-05	2.22E-05
28	chicken, meatless	0.0138	0.5	25	4.62E-04	5.33E-07	4.70E-07
29	frankfurter, meatless	0.004	0.5	50	4.62E-04	3.09E-07	2.72E-07
30	luncheon slices, meatless	0.0088	0.5	50	4.62E-04	6.80E-07	5.99E-07
31	meatballs, meatless	0.0129	0.5	50	4.62E-04	9.97E-07	8.78E-07
32	soy burger	0.0154	0.5	50	4.62E-04	1.19E-06	1.05E-06
33	vegetarian meatloaf	0.0055	0.5	25	4.62E-04	2.13E-07	1.87E-07
34	soy burger with cheese	0.0067	0.5	25	4.62E-04	2.59E-07	2.28E-07
35	teriyaki sauce	0	0.5	25	4.62E-04	0.00E+00	0.00E+00
36	turnover, meat filled	0.0293	0.5	25	4.62E-04	1.13E-06	9.97E-07
37	croissant, meat filled	1.2875	0.5	25	4.62E-04	4.98E-05	4.38E-05
38	Mexican dinner (including beef enchilada, beef burrito and cheese enchilada)	0.0977	0.5	25	4.62E-04	3.78E-06	3.33E-06
39	Mexican dinner with chicken	0	0.5	25	4.62E-04	0.00E+00	0.00E+00
40	chicken and noodles dinner	0.0242	0.5	25	4.62E-04	9.35E-07	8.24E-07
41	beef and grain based soups	0.6525	0.5	50	4.62E-04	5.04E-05	4.44E-05
42	chicken noodle soup	7.0405	0.5	25	4.62E-04	2.72E-04	2.40E-04
43	chicken and rice soups	1.3581	0.5	25	4.62E-04	5.25E-05	4.62E-05
44	turkey noodle soup	0.1682	0.5	25	4.62E-04	6.50E-06	5.73E-06
45	instant soup	1.0482	0.5	75	4.62E-04	1.22E-04	1.07E-04
46	sopa seca – dry soups	*	0.5	25	4.62E-04	0.00E+00	0.00E+00
47	BBQ sauce	0.2161	0.5	25	4.62E-04	8.35E-06	7.36E-06
48	tomato with beef soup	0.0464	0.5	25	4.62E-04	1.79E-06	1.58E-06
49	onion soup, dry mix	0.0196	0.5	25	4.62E-04	7.58E-07	6.67E-07
50	vegetable soups (with beef, chicken, turkey, rice or noodles)	5.0883	0.5	25	4.62E-04	1.97E-04	1.73E-04
51	bacon dressings (regular)	0.0032	0.5	25	4.62E-04	1.24E-07	1.09E-07
52	bacon dressings (low calorie)	0	0.5	25	4.62E-04	0.00E+00	0.00E+00

TOTAL   1.62E-03   1.43E-03   ng/d

               2.70E-05   2.38E-05   ng/kg/d

#6000 lbs of samples 1 & 2 in 13 million lbs of total PF used

A comparison can be made between exposure to PHAs present naturally in foods and PHAs from process flavors. The PHAs that were identified as above the limit of detection (50 ppb) in process flavors were IQ and MeIQx and were therefore used as representative samples of maximum levels of exposure to PHAs from process flavors. Table 3 summarizes the daily *per capita* intakes of IQ and MeIQx from process flavors and from cooked meats.

From Table 3 it can be seen that levels of PHAs from these food sources are extremely low and upon comparison, the PHA intake from process flavors is negligible compared to PHA intake from cooked meats. PHAs were found at trace levels slightly above the limit of detection (50 ppb) in only two out of 102 PF samples. Because process flavors are added to foods at low levels, this estimation of intake of PHAs from process flavors is a conservative estimate and tends to overestimate the potential exposure to PHAs from process flavors.

Conclusions
These studies employed refinements in analytical methods that allowed the accurate analysis of process flavors for the presence of PHAs. Previous efforts were hampered by the inherently low levels of PHAs in foods in general, and by the lack of a reliable and accurate analytical method. Since the studies reported here were conducted, further refinements have been achieved (e.g. Solyakov et al., 1999; Richling et al., 1999) that permit even more accurate assessments of the presence of PHAs in process flavors. However, the more recent refinements do not significantly change the conclusions from the studies reported here.

The studies reported here employed a verified, accurate analytical method and demonstrated that process flavors are not likely to be a significant source of PHAs in the human diet. In this study, 90 out of 102 process flavor samples contained no detectable PHAs while only two samples contained trace levels of a single PHA above the limit of detection (50 ppb). Ten samples contained trace levels (<50 ppb) of PHAs, with usually only one PHA per sample.

The available data indicate that the levels of intake of naturally occurring PHAs from foods, including cooked meats, are approximately 100,000-fold lower than doses used in animal studies in which PHAs demonstrated carcinogenic potential. Because the intake of PHAs from process flavors added to food is far lower than the intake of naturally occurring PHAs from foods, it is highly unlikely that the use of process flavors would pose a significant health risk to humans.

Table 3   Estimated *Per Capita* Intake of IQ and MeIQx From PFs and Cooked Meats

HCA	Intake from PF Use (ng/kg bw/d)	Intake from Cooked Meats (ng/kg bw/d)	
		Minimum	Maximum
IQ	2.70×10^{-5}	0.36	150
MeIQx	2.38×10^{-5}	8.9	33

The FEMA Expert Panel reviewed the data generated in this study, together with other relevant data, and concluded that process flavors do not present a safety concern under current conditions of use (Newberne et al., 2000).

References

Adamson R.H., Takayama S., Sugimura T., and Thorgeirsson U.P. Indicution of hepatocellular carcinoma in nonhuman primates by the food mutagen 2-amino-3-methylimidazo-[4,5-f]quinoline. Environmental Health Perspectives. 102, 190. 1994.

Alldrick A.J. and Rowland I.R. Distribution of radiolabeled IQ and MeIQX in the mouse. Toxicology Letters. 44, 183. 1988.

Bergman K. Autoradiographic distribution of 14C-labeled 3H-imidazo[4,5-f]quinoline-2-amines in mice. Cancer Research. 45,1351. 1985.

De Meester C. Bacterial mutagenicity of heterocyclic amines found in heat processed food. Mutation Research. 221, 235. 1989.

Gooderham N.J., Soames A., Rice J.C., Boobis A.R. and Davies D.S. Distribution and elimination of [2-14C]amino-3,8-dimethylimidazo[4,5-f] quinoxaline in mice. Human and Experimental Toxicology. 10, 337. 1991

Grivas S., Nyhammar T., Olsson K. and Jagerstad K. Formation of a new mutagenic diMeIQx compound in a model system by heating creatinine, alanine and fructose. Mutation Research. 151, 177. 1985.

Gross G.A. Simple methods for quantifying mutagenic heterocyclic aromatic amines in food products. Carcinogenesis. 11, 1597. 1990.

Ito N., Hasegawa R., Sano M., Tamano S., Esumi H., Takayama S. and Sugimura T. A new colon and mammary carcinogen in cooked food, 2-amino-1-methyl-6-phenylimidazo[4,5b]pyridine (PHIP). Carcinogenesis. 12(8), 1503. 1991.

Janiec K. and Manley C.H. Regulating process flavors. Presentation to the Division of Agricultural and Food Chemistry of the American Chemical Society. 7 September 2003.

Kato T., Ohgaki H., Hasegawa H., Sato S., Takayama S., and Sugimura T. Carcinogenicity in rats of a mutagenic compound, 2-amino3,8-dimethylimidazo[4,5-f] quinoxaline. Carcinogenesis. 9, 71. 1988.

Kato T., Migita H., Ohgaki H., Hasegawa H., Sato S., Takayama S., and Sugimura T. Induction of tumors in the Zymbal gland, oral cavity, colon, skin and mammary gland of F344 rats by a mutagenic compound 2-amino-3,4-dimethylimidazo[4,5-f] quinoline. Carcinogenesis. 10, 601. 1989.

Keating G.A. and Bogen K.T. Methods for estimating hetrocyclic amine concentrations in cooked meats in the U.S. diet. Food and Chemical Toxicology. 39, 29. 2001.

Knize M. and Salmon C. Heterocyclic amine analysis of process flavors. Unpublished report submitted to FEMA. 1998.

Manley C.M. Overview of the commercial production and chemistry of process flavors. Proceedings of the Toxicology Forum Summer 1999 Meeting. 1999.

Matsukura N., Kawachi T., Morino K. Phgaki H. and Sugimura T. Carcinogenicity in mice of mutagenic compounds from a tryptophan pyrolyzate. Science. 213, 346. 1981.

Munro I.C., Kennepohl E., Erikson R.E., Portoghese P.S., Wagner B.M., Easterday O.D., and Manley C.M. Safety assessment of ingested hetrocyclic amines. Regulatory Toxicology and Pharmacology. 17(2), S1. 1993.

Newberne P., Smith R.L., Doull J., Feron V.J., Goodman J.I., Munro I.C., Portoghese P.S., Waddell W.J., Wagner B.M., Weil C., Adams T.A., and Hallagan J.B. GRAS flavoring substances – 19. Food Technology. 54(6), 66. 2000.

Ohgaki H., Kusama K., Matsukura N., Morino K. Hasegawa H., Sato S., Takayama S., and Sugimura T. Carcinogenicity in mice of a mutagenic compound, 2-amino-3,4-dimethylimidazo[4,5-F] quinoline from broiled sardine, cooked beef and beef extract. Carcinogenesis. 5, 921. 1984a.

Ohgaki H., Matsukura N., Morino K., Kawachi T., Sugimura T and Takayama S. Carcinogenicity in mice of a mutagenic compounds from glutamic acid and soybean globulin pyrolysates. Carcinogenesis. 5, 815. 1984b.

Ohgaki H., Hasegawa H., Suenaga M., Sato S., Takayama S. and Sugimura T. Induction of tumors in the forestomach and liver of mice by feeding 2-amino-3,4-dimethylimidazo[4,5-f] quinoline (MeIQ). Proceedings of the Japanese Academy of Sciences. 61, 137. 1985.

Ohgaki H., Hasegawa H., Suenaga M., Sato S., Takayama S. and Sugimura T. Carcinogenicity in mice of a mutagenic compound 2-amino-3,8-

dimethylinidazo[4,5-f] quinoline (MeIQx) from cooked foods. Carcinogenesis. 8, 665. 1987.

Richling E., Kleinschnitz, M. and Schreier P. Analysis of heterocyclic aromatic amines by high resolution gas chromatography-mass spectrometry: a suitable technique for the routine control of food and process flavors. European Food Research and Technology. 210, 68. 1999.

Shioya M., Wakabayashi K., Sato S., Nagao M. and Sugimura T. Formation of a mutagen, 2-amino-1-methyl-6-phenylimidazo[4,5-b]-pyridine (PhIP) in cooked beef by heating a mixture containing creatinine, phenylalanine and glucose. Mutation Research. 191, 133. 1987

Sinha R., Rothman N., Salmon C.P., Knize M.G., Brown E.D., Swanson C.A., Rhodes D., Rossi S., Felton J.S. and Levander O.A. Heterocyclic amine content in beef cooked by different methods to varying degrees of doneness and gravy made from meat drippings. Food Chemical Toxicology. 36, 279. 1998a.

Sinha R., Knize M.G., Salmon C.P., Brown E.D., Rhodes D., Felton J.S., Levander O.A. and Rothman N., Heterocyclic amine content of pork cooked by different methods to varying degrees of doneness and gravy made from meat drippings. Food Chemical Toxicology. 36, 289. 1998b.

Skog K.I., Johansson A.E. and Jagerstad M.I. Carcinogenic heterocyclic amines in model systems and cooked foods: a review on formation, occurrence and intake. Food and Chemical Toxicology. 36, 879. 1998.

Solyakov A., Skog K. and Jagerstad M. Heterocyclic amines in process flavors, process flavor ingredients, bouillon concentrates and a pan residue. Food and Chemical Toxicology. 37, 1. 1999.

Sugimura T. Carcinogenicity of mutagenic heterocyclic amines formed during the cooking process. Mutation Research. 150, 33. 1985.

Takayama S., Masuda M., Mogami M., Ohgaki H., Sato S., and Sugimura T. Induction of cancers in the intestine, liver and various other organs of rats by feeding mutagens from glutamic acid. Japanese Journal of Cancer Research (Gann). 75, 207. 1984.

Takayama S., Nakatsuru Y., Ohgaki H., Sato S., and Sugimura T. Induction of cancers in the intestine, liver and various other organs of rats by feeding mutagens from glutamic acid. Japanese Journal of Cancer Research (Gann). 76, 815-817. 1985.

Chapter 3

Regulating Process Flavor

Kimberly Janiec and Charles H. Manley*

Takasago International Corporation, 4 Volvo Drive, Rockleigh, NJ 07647

The practice of making savory flavors involves the use of many types of ingredients and processing conditions. The diversity of these flavors and their processing parameters makes it difficult to easily define the flavor and also to establish their safety in use. The flavor industry has created guidelines for their production and established studies to prove the safety of these materials as food flavorings. The guidelines and regulatory controls for these types for flavors will be discussed in this chapter.

Introduction

The cooking of food materials is a very old method for the preparation of food and for the creation of flavor. Heat creates aroma and taste components for foodstuffs where none existed previously. The science of that transformation is reported in many books and research articles and this knowledge will be enhanced with the information in this chapter. The scientific basis of the chemistry of the thermal development of these flavors is of great interest to the flavor industry and serves as a source of knowledge in building flavors that are commercially useful. These flavors are known as "process flavors" or in some texts "reaction flavors". Their creation follows the age-old result of cooking. Therefore, the relationships of the end products of the "reaction" to well known foods that have been eaten for centuries are most important. To create a roast beef or a grilled chicken flavor is the quest of the flavor industry, using the scientific knowledge of reactions that occur when foodstuffs are heated.

Although the main focus is to have the flavor mimic the aroma developed during the cooking of food, the major goal is also to manufacture the flavor in a way that conforms to safe manufacturing practices as well as to develop flavors that are considered safe for human consumption. It is the foremost priority of the flavor industry to ensure that all the materials used in flavors are safe as food ingredients and it is the responsible of government to create regulations that ensures that safety.

The flavor industry has developed different procedures around the world to evaluate safety and have worked with regulatory groups to ensure that they develop regulations that are scientifically correct, while at the same time remaining open to accept new innovations from the science. Many of the substances used in the industry are pure, chemically defined substances or natural materials extracted by some physical method from various materials historically used as, or in, foods. In the United States these substances can be found in the so called "GRAS" (Generally Recognized As Safe) list that both the Food and Drug Administration (FDA) and the Flavor Extract Manufacturers Association (FEMA) have developed in the years since the enactment of the 1958 Food Additives Amendment of the Pure Food Act (1). Currently the FEMA GRAS list contains over 3000 chemically defined substances and natural materials. The European Union is currently involved in establishing a similar list of approved substances based on a scientific review of the substances and their intended use.

A flavor produced by thermal processing techniques presents a problem in the determination of safety as the end result is not a chemically defined substance, nor is it a simple naturally occurring mixture from food materials such as orange oil, for example. Process flavors are closely related to food, as they are complex mixtures wherein the flavor is developed by a heat process given to the beginning mixture. Science has yet to give us a method to ensure the safety of the food we eat. Indeed, many of the conflicting reports on food and negative health effects only serve to confuse the consumer and may not, in most cases, do anything to protect them. To discuss the regulatory role in the "process flavor" area one must build on the history of the use of the products, their intended use, and the indicators that may be used to ensure their safe use in foods.

The History of Process Flavors

The first "process flavors" were flavors that were developed when meat, soybeans, grains and other foodstuffs were heated, roasted, cooked or exposed in any way to heat. The desirability of cooked food is of major human interest. Our diet has been replete with flavors that have had their creation at the hands of man. Cooking food is, after all, a human function only. In man's search for these types of flavors, he started to use more innovative ways to produce the flavors. The first flavor to be used that was more a flavor invention rather than food was the Hydrolyzed Vegetable Protein (HVP). This material has been in man's diet for more than 150 years. The invention goes back to the Napoleonic times when meat was precious and armies moved on their stomachs. The credit can be given to Liebig of Germany and Maggi of Switzerland for the creation of a flavor by a technique that uses mineral acid to hydrolyze vegetable proteins down to their basic amino acids, and then by heat, reacts into the form flavor (2). The reaction mixture contains flavor, but also a high amount of salt, amino acids and their salts (a major one is glutamate) and color. These mixtures were useful as flavors, but also for their flavor enhancing effects in food. HVP's have been used for 150 years and considered safe by their record of use.

The first patent for a "process flavor" may be considered one that Dr. C. G. May of Unilever received in 1960 (3). His group's invention used the knowledge of basic reactions between amino acids and sugars to develop a strong meat-like character. After that work, and with the advent of gas chromatography and the mass spectrometer, it became a focus of the flavor industry to create more and improved flavors based on these concepts and the increased scientific knowledge related to precursors and heat. This is the purpose of this book and the many others that have preceded it.

Regulatory Toxicology and Food Flavors

The primary consideration of the FDA in the area of flavors is their safety for their intended use for human consumption. Although many would like a perfect world where everything is absolutely safe, the FDA along with the laws that define food additives and GRAS substances that can be used in food recognizes that the definition of safe must be based on the reasonable certainty that no harm will result in the material's use in the food supply. Section 409 of the Federal Food, Drug and Cosmetic Act requires that a food additive be shown to be safe under its intended conditions of use before it is allowed in foods. A food additive therefore must meet the government's (FDA) requirements for safety prior to marketing the material. However, in the 1958 Food Additive Amendment, the United States Congress recognized that it would be impractical to require safety testing of the large numbers of ingredients that were in use prior to 1958. Therefore, Congress specified that "generally recognized as safe" substances were exempt from the pre-market review process.

According to the food additive procedural regulations established by the FDA (21 CFR 170.30), general recognition of safety may be based only on the evaluation of the safety of substances added to food by experts qualified by scientific training and experience. The government, and the Act, did not indicate whose experts should evaluate safety, only that experts qualified to make the judgments be used. Based on this, FEMA established their Expert Panel to review the safety of substances used by the flavor industry. In doing so they established the well used FEMA GRAS list, starting in the late 1960's with the publication of GRAS III in "Food Technology" (4). Experts, including input from FEMA's Expert Panel, developed methods of safety evaluation. FEMA statements regarding their Expert Panel reviews can be found in a number of papers (5). Because of the complexity of "process flavors" FEMA had to devise a different method then that used to establish GRAS status for simple natural extracts and chemically defined substance. Their method evolved over a period of years and included a significant investment of resources by both FEMA and the flavor industry.

More on GRAS

The FDA on June 25, 1971 revised 21 CFR 170.30 to be more specific about "GRAS." The revised version declared that any substance of natural biological origin (including those modified by conventional processing) and consumed primarily for nutrient properties before January 1, 1958 without detrimental effect, would be considered GRAS with no need for official notice. It also listed five categories for GRAS classification, with the understanding that experts must also provide convincing evidence of their safety:

- Substances modified by processes (after January 1, 1958) that may reasonably be expected to significantly alter the composition of the substance.
- Substances that had been significantly altered by breeding or selection and the change would be considered a significant change to the nutritive value or toxic properties of the constituents.
- Distillates, isolates, extracts, concentrates of extracts or reaction products of substances considered as GRAS.

- Substances not of natural biological origin including those for which evidence is offered that they are identical with a GRAS counterpart of natural biological origin.
- Substances of natural biological origin intended for consumption for other than their nutrient properties.

A further refinement of the GRAS regulations indicated that the general recognition of safety through scientific procedures must ordinarily be based on published literature, and must be of the same scientific quality to that of a food additive application.

The FDA has indicated that processed flavors were not recognized prior to September 6, 1958 and therefore are not prior-sanctioned GRAS materials. The statute authorizes the scientific community, and not the FDA, to make judgments of GRAS. Therefore, the FDA assumes that the manufacturer has scientific evidence of the safety of process flavors.

Although the FDA has never legally defined process flavors as GRAS, representatives of the FDA have publicly stated that based on all available information, process flavors meet the requirements for GRAS status (6). The FDA considers that these flavors are GRAS based on the following:

- The manufacturing process is related to high temperature cooking.
- There is a selection of ingredients similar to the preparation of gravy. The belief is that process flavors are produced on the selection of natural ingredients that have been found to create a high flavor note when cooked at a high temperature similar to gravy.
- The use level of process flavors is low, as is true of all flavors; therefore the consumption rate is low in consumers' diets.

These conclusions were made on the presentation of scientific, manufacturing and use information provided by the manufacturers though their trade association, the Flavor and Extract Manufacturers Association (FEMA) of the United States (7).

FEMA's Development of Methods for Reviewing the Safety of Process Flavors

During the late 1970s the FDA was asked by President Richard Nixon to make a full review of GRAS materials used in food at that time. This review was initiated in light of the de-listing of Cyclamate as a potential carcinogen. HVPs and Autolyzed Yeast Extracts (AYE) were two GRAS materials that were reviewed by a special panel called the Selected Committee on GRAS Substances (SCOGS). The HVP/Yeast industry established an industry group, the International Hydrolyzed Protein Council (IHPC) to develop data to support the safety of HVPs and AYEs. A few companies already had significant safety data on HVP and they, in turn, provided this data to the IHPC. One company's data on safety was related to "process flavors" based on HVPs and a mixture of amino acids and reducing sugars. The SCOGS panel accepted the safety data, but questioned that the material was not the description of what was given for HVPs. At that time the IHPC turn the problem of "process flavor" over to FEMA and FEMA established a Committee to develop information about the practice of making "process flavors".

During the late 1970's and early 1980's the FEMA Committee established a number of pieces of data valuable in the determination of the safe use of "process flavors". They included:
- Review the practice of the industry by a survey of methods, precursors, reactions conditions (including pH and time/temperature) and amounts produced.
- Development of a profile of the manufacture and ingredients used in creating the commercial flavor.
- Establish a program to educate the FEMA Expert Panel on the practice of the industry and the sharing with that group of safety data developed by individual flavor companies.
- This led to the establishment of international guidelines by the International Organization of Flavor Industries (IOFI).

The result of this program led to the conclusion that "process flavors" were not a safety concern at their current use levels and manufacturing conditions. Further, there has been no scientific data generated by, or known by the FDA, that challenges the industry's assertion that "process flavors" are safe. However, certain substances may be present in "process flavors" that may be considered toxic. The substances in question have been the subject of significant research and risk assessments with the resulting conclusion of no health concerns. One of these studies is reported in this text (8).

Taking into consideration this safety concern of process flavors, IOFI developed international guidelines for the manufacturer of "process flavorings". These guidelines can be found in Appendix I of this chapter. The Council of Europe adopted these guidelines as pseudo regulatory for the European countries and more recently the European Union (EU) has put the guideline into a flavor directive (88/388/EEC) as noted in Table I. Therefore, in the EU there is a legal flavor category called "process flavoring" whereas in the United States for FDA controlled foods there is no distinction made between flavors and "process flavors".

The European Union Definition of Process Flavouring (88/388/EEC) is as follows: a product which is obtained according to good manufacturing practices by heating to a temperature not exceeding 180° C for a period not exceeding 15 minutes. A mixture of ingredients where they may not necessarily have flavoring properties but at least one contains nitrogen (amino) and another is a reducing sugar.

A summary of some of the regulatory controls for process flavors is listed in Table I.

Meat and Poultry Product Regulations

In the United States, the FDA is the lead regulatory agency for flavor regulation. FDA regulated products do not need pre-market clearance by the agency; however under the Meat and Poultry Inspection Act of 1906, the U. S. Department of Agriculture (USDA) is the lead agency for meat and poultry products. By law the USDA must approve all materials used in meat and poultry prior to marketing. To help the industry in marketing their flavors and seasoning blends, the USDA, under their inspection arm (Food Safe Inspection Service), established a formula review service in Washington D.C. This service, known as the "Proprietary Mix Committee" (PMC), reviewed and approved the labeling of flavors and seasoning blends and issued a so-called "PMC letter" that can be used by companies to show local USDA food plant inspectors that the flavor or seasoning blend was approved for use in a USDA inspected product. This procedure protected the formulation (company's trade secret) from review by local inspectors.

The development of "process flavors" based on proteins and other materials considered to have nutritional content or functional properties other than flavor became a major problem for the PMC. Their charge and professional view is that all these materials should be labeled. Labeling of flavor ingredients was exempted under a FDA decision. However, the USDA PMC considered them non-flavor ingredients. This situation led to havoc in the flavor industry as companies had two labels for one flavor and in many instances a customer company would indicate the use of the flavor in a USDA product after being given an FDA label with disclosure of the protein or nutritional based materials used in making the "process flavor".

Table I, global status of process flavors

Country	USA	European Union	Japan
Title	Flavor	Process Flavor	Flavor
General Restrictions	Considered GRAS per 21 CFR Part 170.30	Must meet guidelines 88/388/EEC and IOFI	Must be safe
Specific Restrictions	Must meet requirements provided in Table 2	Only amino acids appearing on EU Register of Flavoring Substances may be used*	L-Proline not permitted
Heterocyclic Amine	No limits	15 µg/kg	No limits
Mono Chloropropanols	No limits	20ppb (liquid basis) 50ppb (dry basis)	No limits
BSE	No Specific Risk Material allowed	No specific regulation	No specific regulation

*Specific country regulations may vary, please consult local authorities.

Industry representatives met with FSIS and PMC professionals to inform them of the use of the ingredients and how they were, indeed, used to produce flavor and not used for other functionality including nutrition. The results of the discussion was a directive that allow certain materials used in "process flavors" to be exempted from labeling if the total mixture call "process flavor" was subject to a heating process. The elements of the directive are shown in Table 2. Still of major concern by both the industry and the USDA and non-exempt from labeling were proteins. Their concern was based both on potential allogenecity and religious orientation. Therefore in the USA, under the USDA there does exist a category of "process flavor", but it is only one that allows companies to self affirm that the flavor's components may be exempt from labeling.. Upon establishment of this procedure the PMC was abandoned and the industry now certifies that their flavor meets the USDA requirements noted in Table II.

Polycyclic Heterocyclic Amines

During the 1980s research on roasted meat and other types of heated protein containing foods indicated the presence of Polycyclic Heterocyclic Amines (PHA). These materials were shown to be mutagenic and major carcinogens. Cooked foods contained these materials; so it was reasoned that "process flavors" would also contain them. Foods and their methods of cooking are not regulated except for

---

. Table II, USDA regulation for ingredients labeling

With the following exceptions and under the conditions described below, ingredients consumed in the reaction may be listed collectively as reaction (process) flavors.

Exceptions:
- All ingredients of animal origin.
- All non-animal proteinaceous substances (including MSG, HVP and AYE)
- Thiamine hydrochloride, salt and complex carbohydrates
- Any other ingredient that is not consumed in the reaction

Conditions:
- Reaction contains amino acid(s), reducing sugar(s), and protein substrates.
- Treated with heat at 100°C or greater for a minimum of 15 minutes.

---

microbiological stability of a product. However, food additives and GRAS materials are regulated when it comes to materials that are known to be hazardous to human health. Therefore, PHAs were of interest to FDA and the knowledge of their presence in "process flavors" was of major concern. This led the industry to establish a study to evaluate the extent to which they occur in "process flavors" and their potential risk in the human diet. A chapter in this text reviews the study's results and the finding that "process flavors" were not a health threat due to PHAs. There is no intention to set limits in the USA because there are no safety concerns.

This also being an issue of concern in Europe, studies were performed and the EU is considering establishing limits on the amount of PHAs found in process flavorings. The proposed limits have been set at 15 µg/kg. Specific names of the PHAs with limits can be found in Table III.

Mono- and Di-Chloropropanols in HVPs

The inclusion of HVPs into "process flavors" brings in a problem that the IHPC and hydrolysate industry has dealt with for many years. During the manufacturing of HVPs, the high acid level (hydrochloric acid) hydrolyzes the

Table III, maximum proposed levels of PHAs in process flavors

Substance	Max. µg/kg
2-amino-3,4,8-trimethylimidiazo [4,5-f] quinoxaline (4,8-DiMeIQx)	15
2-amino-1-methyl-6-phenylimidiazol [4,5-b] pyridine (PhIP)	15

residual fat in the protein and the glycerin formed from hydrolysis is chlorinated to form trace amounts of mono- and di-chloropropanols. These materials are known as strong reproductive toxins. The manufacturers have been able to modify their processing to eliminate the formation of these substances. The FDA has accepted that HVPs are processed under these procedures and have not established a residual limit on them. In Europe (EC Directive 466/2001) a limit has been set for mono dichloro compound only, which is 20 ppb (liquid basis) and 50ppb (dry basis).

Other ingredient labeling issues

Allergens

The issue of labeling of allergens has become a recent concern to both governing agencies as well as consumers. Proteins from various sources are utilized in the creation of "process flavors", some of which may be considered to be allergens, and therefore should be labeled. The materials that are considered to be allergens in their respective regions can be found in Table IV.

Table IV, definition of allergenic materials

USA	Shellfish and shellfish products, Cereals and cereal products, eggs and egg products, fish and fish products, dairy and dairy products, peanuts and peanut products, soybeans and soybean products, tree nuts and tree nut products, and sulfites (only when ≥ 10ppm)
EU	Cereals containing gluten, fish, crustaceans, egg, peanut, soy, milk and dairy products including lactose, nuts, celery, mustard, sesame seed and sulfites (only when ≥ 10ppm)
Japan	Mandatory labeling required of: Buckwheat, eggs, milk and milk products (lactose included), peanut and wheat. Recommended labeling required by notice of MHLW: Abalone, crab, mackerel, salmon, salmon roe, shrimp, squid, beef, chicken, pork, gelatin, apple, kiwi, orange, peach, matsutake .mushroom, soybean, yam and walnut. Additionally, discussions are currently underway to declare banana as an allergen.

While most governments are currently addressing the issue of allergens through legislation (in the USA; Food Allergen Labeling and in the EU; Consumer Protection Act of 2003/EU amendment to Directive 2000/13/EC), until such laws are in effect, the flavor industry should be obligated to divulge to customers any inclusion of proteins from allergenic sources. An important note here is that current scientific evidence does not allow for a determination of intake thresholds, and therefore the presence of any proteins, regardless of the amount, from the above sources requires labeling.

Bovine spongiform encephalopathy (BSE)

With the recent case of BSE in the United States, along with past cases in both Europe and Japan, there has been sparked concern over the presence of potentially harmful protenacious animal material in flavors. Since a protein nitrogen source (from an animal source) is a typical component of a "process flavor", it is an obvious concern. The USDA has recently filed an interim file rule (9 CFR Part 301, 309, et al.) prohibiting the use of Specific Risk Material in food. Specific Risk Material is defined as the skull, brain, trigeminal ganglia, eyes, vertebral column, spinal cord and dorsal root ganglia of cattle 30 months of age or older and the small intestine of all cattle. It is the responsibility of each flavor manufacturer to ensure that the protein nitrogen source used in the manufacture of a process flavor does not contain any of these prohibited materials. In Europe, in order to eliminate risk, flavor manufactures ensure that amino acids are not derived from any beef material. In Japan, there are no specific restrictions.

Genetically Modified Organisms (GMO) issues

Europe has recently passed legislation regarding the labeling and traceability of GMO and products produced from GMOs (Regulation 2003/1829EEC and 2003/1830). The intent of this regulation is to ensure traceability of GMOs from farm to shelf, as well as provide consumers with the information necessary for them determine a product's status, and then choose for themselves whether to include or exclude these products from their diets. Hydrolyzed Vegetable Proteins (HVP's) are common ingredients in process flavors; corn and soy are ypical of these. At the same time, corn and soy are among the most widespread vegetables being genetically modified today. It's clear then that flavors utilizing them would be of a concern for the presence of GMOs. Therefore, traceability and labeling of these in any foodstuffs is required. In addition, you must declare the presence of GMO corn and soy in all flavors, including "process flavors". In the United States no such regulations are in place, nor are any in place in Japan at this time.

Conclusions

Process flavors represent a very complex and diverse groups of flavoring materials that are regulated in various ways around the world. Because of this complexity it is important that there is a full understanding of guidelines of manufacture, the justification of safety and the rules of labeling. If you have any questions you should consult your country's flavor trade organization or the International Organization of Flavor Industries.

Appendix 1

IOFI Code of Practice

Definition: A thermal process flavouring is a product prepared for its flavouring properties by heating food ingredients and/or ingredients which are permitted for use in foodstuffs or in process flavorings.

4.   **Production of process flavourings**
- Process flavourings shall comply with national legislation and shall also conform to the following:

4.1.   Raw materials for process flavourings
- Raw materials for process flavourings shall consist of one or more of the following:

4.1.1.   A protein nitrogen source:
-protein nitrogen containing foods (meat, poultry, eggs, dairy products, fish, seafood, cereals, vegetable products, fruits, yeasts) and their extracts
-hydrolysis products of the above, autolyzed yeasts, peptides, amino acids and/or their salts

4.1.2.   A carbohydrate source:
-foods containing carbohydrates (cereals, vegetable products and fruits) and their extracts
-mono-, di- and polysaccharides (sugars, dextrins, starches and edible gums)
-hydrolysis products of the above

4.1.3.   A fat or fatty acid source:
-foods containing fats and oils
-edible fats and oil from animal, marine or vegetable origin
-hydrogenated, tranesterfied and/or fractionated fats and oils
-hydrolysis products of the above

4.1.4.   Ingredients listed in Table 1

4.2.   Ingredients of process flavourings

4.2.1.   Natural flavourings, natural and nature-identical flavouring substances and flavour enhancers as defined in the IOFI Code of Practice for the Flavour Industry.

4.2.2.   Process flavor adjuncts
- Suitable carriers, antioxidants, preserving agents, emulsifiers, stabilizers and anticaking agents listed in the lists of flavour adjuncts in Annex II of the IOFI Code of Practice for the Flavour Industry.

4.3.   Preparation of process flavourings
- Process flavourings are prepared by processing together raw materials listed under 4.1.1 and 4.1.2 with the possible addition of one or more of the materials listed under 4.1.3 and 4.1.4

4.3.1.   The product temperature during processing shall not exceed 180°C

4.3.2.   The processing time shall not exceed ¼ hour at 180°C, with correspondingly longer times at lower temperatures.

54

4.3.3. The pH during processing shall not exceed 8.
4.3.4. Flavourings, flavouring substances and flavour enhancers (4.2.1) and process flavour adjuncts (4.2.2) shall only be added after processing is completed.

4.4. General Requirements for process flavourings
4.4.1. Process flavourings shall be prepared in accordance with the General Principles of Food Hygiene (CAC/VOL A-Ed.2(1985)) recommended by the Codex Alimentarius Commission.
4.4.2. The restrictive list of natural and nature-identical flavouring substances of the IOFI Code of Practice for the Flavour Industry applies also to process flavourings.

5. Labelling
The labeling of process flavourings shall comply with national legislation.
5.1. Adequate information shall be provided to enable the food manufacturer to observe the legal requirements for his products.
5.2. The name and address of the manufacturer or the distributor of the process flavouring shall be shown on the label.
5.3. Process flavour adjuncts have to be declared only in the case they have a technological function in the finished food.

Materials Allowed In Processing

Herbs and spices and their extracts, Water, Thiamine and its hydrochloric acid salt
Ascorbic Acid, Citric Acid, Lactic Acid, Fumaric Acid, Malic Acid, Succinic Acid, Tartaric Acid
The sodium, potassium, calcium, magnesium and ammonium salts of the above acids
Guanylic acid and Inosinic acid and its sodium, potassium and calcium salts
Inositol, Sodium, Potassium-and Ammoniumsulfides, hydrosulfides and polysulfides, Lecithine
Acids, bases and salts as pH regulators: Acetic acid, hydrochloric acid, phosphoric acid, sulfuric acid
Sodium, potassium, calcium and ammonium hydroxide
The salts of the above acids and bases;
Polymethylsiloxane as antifoaming agent (not participating in the process).

References

(1) Code of Federal Regulations, Title 21 section 172.510 to 172.590
(2) Manley, C. H., In Thermal Generation of Aromas; Editors, T. H. Parliament, R. J. McGorrin and C-T Ho, American Chemical Society, Washington DC, 1989, pgs 12-21.
(3) May, C. G., Akroyd, P., British Patent 655,350, 1960.
(4) Hall, R. L., Oser B. L. Food Tech. 1965: 151-197.
(5) Newberne, P., Smith, R. L., Doull, J.,Goodman, J. I., Munron, I.C., Portoghese, P.S., Wagner, B.M.,Weil, B.M.,Woods,L.A.,Adams. T.B.,Hallagan,J.B.,Ford,R.A., Food Tech., 1998: 65-92.
(6) Lin, L. J. In Thermally Generated Flavors; Editors, T. H. Parliament, J. J. Morello and R. J. McGorrin, American Chemical Society, Washington DC, 1995, pgs 7-15.

(7) Newberne, P., Smith, R. L., Doull, J.,Goodman, J. I., Munron, I.C., Portoghese,P.S.,Waddell,Wagner,B.M.,W.J.,Weil,Adams. T.B.,Hallagan, Food Tech., 2000, 66-84.
(8) Hallagan, J. In Savory Flavors; Editor, D.K.Weerasinghe, American Chemical Society, 2003, (This text pg?).

Chapter 4

An Introduction to Kosher and Halal Issues in Process and Reaction Flavors

Edith Cullen-Innis

Product Safety and Regulatory Affairs, Firmenich Inc.,
250 Plainsboro Road, Plainsboro, NJ 08536

Introduction

In the flavor business, worldwide, Kosher and Halal certification plays an important role. There are several international kosher and halal supervisory agencies who work closely with those in the food business to provide reliable guarantees to the consumer by certifying hundreds of thousands of products each year.

The Orthodox Union (the OU) supervises manufacturing plants worldwide, supervising products in 68 countries across the globe. While there are several kosher supervisory agencies, the OU by far is the largest worldwide, and supervises the majority of the food industry

Much like kosher, there are several halal supervisory agencies worldwide. The more conservative agencies, however, find a greater acceptance by the halal consumers worldwide. IFANCA (Islamic Food and Nutrition Council of America) in the United States, the IFCE (Islamic Food Council of Europe) in Europe, and the MUI (Majelis Ulama Indonesia), in Jakarta, all supervise halal production. The staff of IFANCA and IFCE and MUI are all highly educated and trained in food sciences and understand very well the rigors of our flavor technology.

Halal applications can be the same as the kosher items, except that no ethyl alcohol is allowed and, of course, no pork products for either kosher or halal.

The regulations that guide the planning and production of Kosher and Halal products will be discussed in this chapter

1. Kosher

In order for a product to be kosher, the ingredients and the equipment must *both* be kosher. The retail product that the consumer sees is clearly marked with the OU logo on the product label, which indicates that this product contains only kosher ingredients and was made only on kosher equipment.

Most companies have a Kosher coordinator on staff and Rabbis from a supervisory agency dedicated to each of the kosher manufacturing sites. They work together to make sure that the kosher logo will be displayed on many products that are requested to be made kosher.

Kosher products cover a wide range of consumer products. In the food industry, beverages like teas and coffees, juices, nutritional supplements and dairy drinks are all available with the approval of the OU. The bakery industry produces cakes, muffins and breads that are kosher. Sweet products like gums, candies, yogurts and puddings are kosher. Cereals, soups, snack foods, etc., can be made as Kosher products. In some instances even toothpaste and cough medicines are available as Kosher products. Any one of these products that contain or require flavoring need to be supplied with Kosher flavors.

There are only a few products in the world which can *never* be kosher: civet, castoreum, carmine (no insects), pork, shellfish (and all fish without fins and scales, such as shark). Also, mixtures of dairy and meat, and mixtures of fish and meat can *never* be kosher.

There are two marketed ingredients which are not available as kosher: natural cognac oil and wine fusel oil. All fruits and vegetables (except from Israel) are kosher, all steam distilled essential oils, spices, inorganic compounds, most FD&C colors, petrochemicals, and all turpentine derivatives are kosher Ingredients that are, or can be derived from, grape are very sensitive for kosher. This impacts juices, vinegars, and some alcohols.

Kosher dietary laws divide all food and their ingredients into four categories: dairy, parve, meat and non-kosher.

Kosher Dairy

The definition of "kosher dairy" means milk and any milk products, like cheese or butter. The kosher dairy product must come from a kosher species of animal in order to be considered kosher. Any milk derivative, like casein or butter distillate, is considered dairy when used in kosher foods, but may not be required to be claimed as dairy by the FDA.

Ingredients that have dairy origin or are derived from cheese are closely monitored. Whey, and all fermentation chemicals using whey, enzyme modified cheeses and butter oils are all considered dairy by a Rabbi.

Dairy ingredients used on manufacturing equipment will render all ingredients processed on the equipment as dairy. If parve or meat production is scheduled to follow a dairy production, then the mandatory 24 hour idle time will be required for the equipment, followed by a "Kosherization" of the equipment. This process will be discussed later in the chapter.

Kosher Parve

Parve means "neutral." Neutral means food that contains neither milk nor meat. Fruits, vegetables and fish are all considered parve. All steam distilled essential oils, pure spices, inorganic compounds, most FD & C colors are considered parve. Most petrochemicals and all turpentine derivatives are kosher parve. Even though grape is considered parve, ingredients that are derived from grape are sensitive for kosher production. This impacts juices, vinegars, and some alcohols. All grape products must be supervised by the Rabbi for production.

All fruits and vegetables from Israel must be carefully reviewed in order to guarantee their kosher status. Special rules apply to products from Israel, and it is best to check with the supervising Rabbi before assuming that, just because the product is parve, it will be acceptable.

Kosher Meat

Foods made from meat or meat by-products are in this category. Only meats from kosher species of animals or fowl are permitted and only if slaughtered under kosher guidelines. In order for an animal to be kosher it must chew its cud and have a split hoof. For example, a pig has a split hoof, but does not chew its cud. A camel chews its cud, but does not have a split hoof. Kosher species include cattle, sheep, chicken and turkey. Following a ritual slaughter by a properly trained specialist, blood must be removed from the meat and properly prepared by soaking and salting.

REACTION FLAVORS

Reaction flavors are of special interest to the Rabbi because this process involves heated equipment. The ingredients often used in reaction (process) flavors could also be of concern. For example, a savory reaction flavor could use meat (which includes chicken fat), amino acids (which can be porcine based) and enzymes (which can also be meat based).

The ingredients, as briefly mentioned earlier in the chapter, must all be kosher in order for your process to be accepted by a Rabbi. There are several amino acids which are kosher, even cysteine from human hair. You can buy kosher chicken fat, kosher lipase, kosher alpha alanine, etc.

Unless you have equipment dedicated to producing Kosher reaction flavors, you must have the equipment "Kosherized" before each use. "Kosherizing" the equipment means boiling water in the vessels to be used, and streaming boiling water through any hoses, tubes or pumps where the product will pass, and also boiling utensils, such as scoops or screens. Boiling means boiling, i.e., 212°F or 100°C to a Rabbi. Prior to kosherizing, the equipment must remain idle (at ambient temperature) for at least 24 hours.

Reaction flavor equipment raises a red flag for the Rabbi, if the heating process in the vessel exceeds 110°F. For example, you can warm up solidified vanilla at a temperature lower than that, and your equipment is not considered "reaction" equipment to a Rabbi.

Why does the equipment need to be "Kosherized?" Metal expands and contracts during heating, so any residue from previous non-kosher productions will, to a Rabbi, still be present. Any non-kosher residue would render a kosher ingredient or kosher compound as non-kosher.

Enzymes always need verifiable ingredient source data or better, a Kosher certificate. Since enzymes are secreted by bacteria and bacteria is grown in a medium, it is possible for this medium to contain non-kosher ingredients, such as animal tissue.

Animal products themselves, such as chicken fat or beef gelatin, must be ordered and stored as Kosher products. Even a droplet of chicken fat will mean that your flavor will be called a Kosher *meat* product. Consider this when developing a good butter flavor for baked goods – this meat designation will follow your flavor all the way to the shelves of the local supermarket.

2. Halal

Halal means permitted or lawful.
Haram means prohibited or unlawful.
Halal and haram are universal terms, not limited to food.
Makrooh means detested or unhealthy.
Mashbooh means questionable or something for which there is a lack of information or a difference of opinion (e.g., cheese, because it may be made with enzymes from pork or calf and should be investigated further before purchasing or eating, or any food products containing gelatin are questionable).

In order for a food or flavor product to be considered halal, both the ingredients and the equipment used to produce a product must be permitted by Muslim law. The mullah is our field inspector and the Imam is our leader or halal supervisor, like the rabbi is for kosher.

The mullah will consider everything halal except: pork, carrion (meat of dead animals), blood, improperly blessed meat, alcohol and intoxicants.

Alcohol

Any potable alcohol, such as ethyl alcohol (ethanol) is the alcohol of concern to halal food or flavor production. Propylene glycol, frequently used in food and drug industries, is not a concern. Alcoholic beverages such as beer and wine or distilled spirits (vodka, whiskey, etc.) are forbidden under halal guidelines.

In the flavor business, alcohol can be present in an ingredient in many ways. We try to analyze possible alcohol presence through several methods. For example, intentionally added ethanol is a zero tolerance ingredient, per all halal guidelines worldwide. I can also review the alcohol content in an orange oil, and if present, may be permitted by halal law since it is intrinsic to the natural fruit oil. Please see the below listing:

1) Intentionally added ethanol, allowed at 0 ppm (i.e.., Wine, distilled spirits)
2) From fermentation, ethanol is allowed at 10 ppm or less (i.e.., from vinegars)
3) Residually occurring, allowed at 10 ppm or less (i.e., from fruit concentrates)
4) Naturally occurring ethanol, allowed at 500 ppm or less (i.e., orange oils)

Certain citrus oils do contain a high degree of intrinsic ethyl alcohol, and must be monitored very carefully, though will not always be forbidden in a flavor application.

It is important to note that, the alcohol issue in Muslim communities differs from region to region, and from country to country.. It is a topic very open to discussion. The best advice I can give you is to know your customer, and know the country where your product will be used.

Alcoholic beverages are forbidden, as we acknowledged earlier. However, the acceptance of ethyl alcohol in your products will depend on the intended use for the your product . Find out if your customer will use your product in a beverage, or a cooked food application.

Alcohol from fermentation will be found in vinegar, but vinegar itself is an acceptable product for halal, so the presence of the fermented alcohol will not be cause to ban the product from use. The only time fermented alcohol will be forbidden above a certain percentage is when found in a potable beverage. The beverage would then be considered an intoxicant

If you know the final application for your product, in a liquid or a dry powder, as long as it is not an "alcoholic drink," then the presence of alcohol - up to 0.1% of a consumer item - is most often allowed. Certain countries, such as Malaysia and Indonesia, may accept less than 0.1%

Fish

Special consideration must be given to fish and seafood . A minority of Muslim groups do not accept fish without scales such as catfish, shark, swordfish , nor shellfish.

Fish proteins are acceptable from whole fish or ground fish.

Dairy and Meat

Milk and eggs of acceptable animal species are permitted. Acceptable species are sheep, goats, lambs, cows, chickens, hens, quails, turkeys. Pork and any porcine derived ingredients are absolutely forbidden.

There are restrictions on some dairy ingredients. For example, enzymes used to make cheese should be of microbial origin to be acceptable as halal. Rennet from calf must be avoided. The active ingredient in rennet is Chymosin. Chymosin can also be produced in the lab, and this biotech product can be halal. Most calf rennet in the US is not halal.

Other enzymes, such as lipase, which could come from pig (which have been used for some Romano cheeses) need to be verified. Even preservatives, such as propionate, must be reviewed before using. Any animal fatty acid derivatives must be carefully monitored, or halal certified by a mullah, before using with other halal ingredients. Animal fatty acids will always be a concern for halal due to possible pork contamination, either of ingredients or from equipment.

When creating cereal and confectionary products, watch for gelatin (sometimes from pork), mono- and di-glycerides, and any other emulsifiers. The emulsifiers are used to prevent oil and water from separating. They can be derived from animal or vegetable sources. Vegetable based mono- and di-glycerides should always be acceptable for halal; animal based products need halal certification.

REACTON FLAVORS

Again, we need to watch for buzz words of the reaction product: **amino acids, enzymes and animal fats**

Heat, in itself, does not present an obstacle for halal manufacturing. When inspecting the reaction equipment, it is important to know the history of your equipment. If, for example, you know that products containing pork or pork by-products were previously used on your equipment you must first clean the equipment thoroughly and then you can produce your product as halal, providing the ingredients are all halal accepted. Once this equipment is cleaned, and the equipment is meant for halal production, you cannot return to using porcine-base ingredients on this equipment.

For beef and poultry products, even using non-halal certified products, you can use the equipment over and over again, as long as the equipment is properly cleaned after each manufacture, using GMP cleaning procedures. For halal, it is recommended that the cleaning water be a temperature higher than the reaction process, or cleaned with acid or alkaline cleaning products, whatever works the best. There is no requirement for a boiling water temperature, as in kosher, and there is no restriction on mineral inorganic cleaners.

Amino acids: the most obvious question to ask is, what is the source of an amino acid? i.e., Cystine/Cysteine can be sourced from human hair or pig bristles. Neither of those sources are acceptable for halal production. Another example is Methionine, often sourced from animal. Alanine, can be found in plants as well as in animal meat. Send your questions to your halal supervisor about the amino acids you intend to use. If you are not a certified producer, go on-line to IFANCA.org, and most information is available at this site

Enzymes: we covered a little in the section on dairy. Again, let me emphasize the need to know the origin of the enzyme you choose to use. Animal derived enzymes will usually not be halal: microbial and plant sources are you safest choices.

Animal Fats: Most often you will be able to source halal chicken meat, and chicken fat. Large orders are easier to procure, but check with your suppliers for the halal designation first.
Beef fat is also available as halal, and pork fat will NEVER be halal.
Before placing your orders, make sure the vendor can provide an acceptable halal certification, acceptable to the manufacturer's halal supervisor and acceptable to the halal agency in the importing country.

HALAL MARKET LOCATIONS

China: 60-80 million Muslims out of 1.6 billion people, Muslims located in the West and Northwest areas of China, only a few thousand Muslims each in Beijing, Shanghai and Taiwan
New Zealand: strong halal activity in small organized Muslim community (2% of population), requires halal certification
Australia: same as New Zealand

Philippines: 5% strong vocal Muslim population and growing
Indonesia: strong Muslim population, must have halal certification
Hong Kong: about 5,000 Muslims only, but educated population with Islamic mosques and Centers and Schools
UK: almost 5% of population and growing, strong and focused halal activity, strong cohesive population, halal certification required
Switzerland: insignificant Muslim population
Lebanon: over 55% of population, reorganization of halal activities underway, requires halal certification
Qatar: almost 95% of population, everything should be halal certified
Kuwait: over 98% of population, everything should be halal certified here also
Saudi Arabia: 100% of population, rigid halal regulations which are part of the Saudi Standards Organization, need good importers here, and Saudi acts as a distributor to region
Dubai: over 96% of population, everything should be halal certified; Dubai airport is a very important community center for Middle East, 80% of migrant workers enter Arab countries here, and most are Muslim, many from India and Pakistan

OTHER POINTS TO CONSIDER FOR HALAL PRODUCTS:

- ➢ Proper labeling with halal symbol or written identification is essential.
- ➢ In the US and Canada, all 6 - 8 million Muslims observe halal. Worldwide, there are about 1.5 billion Muslims, practically all observe halal.
- ➢ Halal food and drinks are for year-round use, not limited to any time period (as Passover for kosher).

To summarize the differences between kosher and halal:

ITEM	KOSHER	HALAL
Definition	A Hebrew term meaning fit or acceptable for Jews	An Arabic term meaning permitted for Muslims
Pork	Prohibited, including all by-products	Prohibited, including all by-products

Animal products and Poultry products	Certain species only permitted, and only if properly slaughtered by trained Jew	Certain species only permitted, and only if properly slaughtered by trained Muslim
Restrictions for meat and poultry	Restrictions on hind quarters, soaking and salting of meat required	Whole carcus in halal, and no soaking or salting is required
Mechanical slaughter	Stunnning or mechanical slaughter is not permitted	Stunning and mechanical slaughter is permitted under certain conditions
Gelatin	Bovine and Fish gelatin only, no porcine gelatin	Bovine and Fish gelatin only, no porcine gelatin
Enzymes, amino acids	Microbial sources are acceptable; human hair cystein is acceptable	Microbial sources are acceptable; human hair cystein is not acceptable
Ethyl Alcohol	Alcohol is acceptable, providing it is not from non-kosher sources, i.e., from non-kosher grape sources	Alcohol as a drink is prohibited; minute amounts of intrinsic alchol is permitted in fruit or processed foods
Fish and seafood	Fish with scales is permitted. Fish without scales and all seafood is prohibited	All fish is permitted. Certain restrictions apply to seafood.
Dairy	Permitted, provided certain regulations in the production are followed	Permitted, provided certain regulations in the production are followed
Eggs	Permitted, providing there is no contamination	Permitted, providing there is no contamination
Plant material	All plants are permitted, providing the restrictions for produce from Israel are followed	All plants are permitted, except any plant material causing intoxication
Combining food groups	Combining meat and dairy, or fish and meat is not kosher.	There are no restrictions.
Manufacturing equipment	All equipment must be sanitized before production, by a Rabbi. 24 hr idle time must be observed following non kosher production.	All equipment must be sanitized, and never shared with pork production.

Special requirements	Passover is a special status with strict guidelines to be followed during the Passover time.	Halal food is halal year-round, with no special requirements.

REAL LIFE APPLICATIONS

Kosher status of retail products is protected in the US by local and states' laws, but there is no legal requirement for import in any country, not even Israel. The biggest consumer population is in the US. Direct sales of kosher foods is a $165 billion industry (Sales for 2002), up from $45 billion in 1996, according to Integrated Marketing Communications, NYC.

Of this consumer population, only 8% said they bought kosher foods for religious reasons.

Halal is recognized by the USDA. There are several countries worldwide at this time that require halal as a legal status for import.

- The Gulf States: Saudia Arabia, Kuwait, Bahrain, Qatar, Oman and UAE states.
- And also countries in South Asia: Indonesia, Malaysia, Singapore, Brunei, Pakistan, Bangladesh, India and Afghanistan

Worldwide there are about 1.5 billion Muslim consumers, from Los Angeles to Indonesia, with the heaviest concentrations being in Northern Africa, the Middle East, and South Asia. There is a growing market in Western Europe (especially France, about 5 million people), and mainly in urban areas, like London and Paris.

There are approximately 8-10 million Muslims living in North America and an estimated 75% of Muslim households follow halal guidelines in one form or another.

According to some reports, it is estimated that spending by halal observers will exceed $15 billion in 2003, in North America alone.

In over 112 countries worldwide, the Muslim consumer base is estimated to be about 1.8 billion people. International halal trade is estimated to be $150 billion per year, (for year 2002 per Canadian Department of Agriculture and US Department of Commerce).

Sources

Kosher Information:

The Orthodox Union (OU)
11 Broadway
New York, New York
212-563-4000

Kashruth, by Rabbi Y. Lipschutz, 1988, Mesorah Publications, Ltd., Brooklyn, NY

Halal Information:

The Islamic Food and Nutrition Council of America (IFANCA)
5901 N. Cicero Avenue
Chicago, Illinois
773-283-3708

The Islamic Food Council of Europe (IFCE)
Rue Bara 150
Brussels, Belgium
32 2524 6733

Halal Food Production, by M. Riaz and M. Chaudry, 2003, crcpress.com
A Muslim Guide to Food Ingredients, by A. Sakr, 1997, Foundation for Islamic Knowledge, Lombard, Ill.

Ingredients and Intermediate Maillard Reactions

Chapter 5

Precursors for Dry-Cured Ham Flavor

Ana I. Carrapiso* and M. Rosa Carrapiso

Tecnología de los Alimentos, Escuela de Ingenierías Agrarias, Universidad
de Extremadura, Ctra. de Cáceres 06071, Badajoz, Spain
*Corresponding author: fax: 00 34 924286201, email: acarrapi@unex.es

Flavor is an outstanding characteristic of dry-cured ham. The
odor-active compounds of this product have been recently
identified, and these data have shown the great importance of
both lipid and amino acid derived compounds. To obtain
samples with odor characteristics similar to those of dry-cured
ham, several temperatures were applied to meat samples, and a
mild temperature was selected. Compounds previously
identified as dry-cured ham odorants were researched by
SPME-GC-MS, and the effect of several factors on them was
checked. The increase of sodium chloride content caused a
general increase in aldehydes, except for 3-methylbutanal. The
addition of sodium nitrite caused a general decrease, specially
in straight-chain aldehydes. The addition of cysteine and
proline on the odorants identified was less important. Reaction
time influenced the odorants, with a marked effect on pentanal
and hexanal.

Introduction

Flavor is one of the most valuable characteristics of dry-cured ham (1, 2), a
valuable non-cooked but ripened pork meat product from southwestern European
countries, and dry-cured ham flavored products are also appreciated. The great
interest of dry-cured ham flavor has lead to an extensive research on this topic.
The first works on the compounds involved in dry-cured ham flavor were

published 40 years ago (reviewed by Flores and co-workers, *1*), although most works appeared in the 90's, mainly focusing on the identification of the volatile compounds and on the influence of different quality factors on them. The works performed to identify the odorants have been much more limited. The first attempt was carried out by Berdagué and co-workers in 1991 and 1993 (*3, 4*), identifying 5 odorants. Flores and co-workers published a list of ham odorants (*5*), and works reporting ham odorants and their relative contribution have been recently published (*6, 7, 8*) (Table 1), about 15 years after the work of Gasser and Grosch on cooked beef (*9*).

Aldehydes are by far the most numerous compounds identified as dry-cured ham odorants, with different odors (green, rancid, toasted) and thresholds in air ranging from 0.09 to 480 ng/L (Table 1). Most of them were identified in the first works focused on dry-cured ham volatile compounds (*1, 2*). Aldehydes are essential for meat flavor (*10*), but large quantities in meat and meat products have been related to lipid oxidation and deterioration (*11*). The effect of several quality factors has been researched and it was found that the rearing system of pigs (*8*) and ripening conditions (*1*) influence on the contribution to odor and the content of some aldehydes.

Sulfur and nitrogen containing compounds are responsible of the meaty notes of dry-cured ham. These low odor threshold compounds (0.0025-280 ng/L in air) (Table 1), and specially heterocyclic compounds, appear at very low concentrations and are unstable, which makes difficult their identification and quantification using the usual gas chromatography-mass spectrometry (GC-MS) procedures. In fact, most of these odorants were not identified in the numerous works performed on dry-cured ham by using GC-MS and they have been recently reported by using gas chromatography-olfactometry (GC-O) (*6, 7, 8*). Sulfur and nitrogen containing compounds are greatly involved in the odor differences between the two main components of ham slices, the fat and the lean (*12*), and some of them seem to be affected by the characteristics of raw meat determined by the rearing system of pigs (*8*). Otherwise, increases in other sulfur-containing compounds such as dimethyl disulfide have been related to microbial spoilage of dry-cured ham (*13*).

Ketones contribute to ham flavor with a wide range of notes and odor thresholds (0.03-1300 ng/L) (Table 1). In dry-cured ham, some of these ketones are affected by the rearing system of pigs (*8*), processing conditions (*1*) or microbial spoilage (*13*).

With regard to acids, odor thresholds are relatively large (except for 3-methylbutanoic acid) and usually appear at such concentrations that allow the identification by usual GC-MS. In fact, a large number of acids has been reported in numerous works focused on ham volatile compounds (*1, 2*), although most of them have not been reported as dry-cured ham odorants. An increase in acids have been reported in spoiled hams (*13*).

Table 1. Odorants found in dry-cured ham (Parma and Iberian ham)[a].

Odorant	Odor description	Threshold in air (ng/L)[b]
Aldehydes		
2-methylpropanal	toasted, fruity	0.97-10[c]
3-methylbutanal	malty	3-6
pentanal	nutty	34-39
hexanal	green	30-53
hex-(Z)-3-enal	leaf-like	0.09-0.36
hex-(E)-2-enal	apple-like	50-480
heptanal	fatty	250
hept-(E)-2-enal	almond-like	52.5-250
octanal	fatty	5.8-13.6
oct-(E)-2-enal	rancid	47
non-(E)-2-enal	fatty	0.1-3.6
2-phenylethanal	honey-like	0.6-1.2
Sulfur-containing compounds		
hydrogen sulfide	rotten eggs	10[c]
methanethiol	rotten eggs	0.02-0.2[c]
2-methyl-3-furanthiol	meat-like	0.0025-0.001
3-mercapto-pentan-2-one	meat-like	0.045-0.18
methional	potato-like	0.1-0.2
2-furfurylthiol	roasty	0.01-280
Nitrogen-containing compounds		
2-acetyl-1-pyrroline	roasty	0.02
2-propionyl-1-pyrroline	roasty	0.02
Ketones		
butane-2,3-dione	buttery	5-30
pent-1-en-3-one	rotten	1-1.3[c]
heptan-2-one	fruity	1300
oct-1-en-3-one	mushroom-like	0.03-1.12
Acids		
acetic acid	acidic, pungent	60
butanoic acid	sweaty	50-2150[c]
3-methylbutanoic acid	sweaty	1.5
phenylacetic acid	honey-like	1000-10000[c]
Esters		
ethyl 2-methylpropanoate	fruity	0.1-0.2
ethyl 2-methylbutanoate	fruity	0.06-0.24
ethyl 3-methylbutanoate	fruity	0.2[c]
Others		
oct-1-en-3-ol	mushroom-like	48
furaneol	caramel-like	1
p-cresol	phenolic	0.3-1
sotolone	seasoning-like	0.015

[a]From reference *1, 7*; [b]From reference *31*; [c]Threshold in water (orthonasal, μg/L).

Esters identified as ham odorants possess fruity odors and low odor thresholds values (0.06-0.24 ng/L) (Table 1) and are branched esters. Quality factors that modify their relative contribution to flavor are the rearing system of pigs (*8*) and ripening conditions (*1*).

Dry-cured ham flavor is clearly different from the flavor of cooked meats and also from the flavor of other non-cooked ripened meats (e.g. sausages or loins). As reported in cooked meat, ham odorants are generated from either lipid-oxidation reactions or reactions involving water soluble compounds such as amino acids, but reactions occur at very different temperatures: usually over 70 °C for cooked meats and less than 35 °C for ripened meats. The main qualitative differences with regard to the odorants of cooked meats are the absence of odorants typically generated at high temperature such as guaiacol and vanillin (*14*). Conversely to other non-cooked but ripened meats, dry-cured hams usually does not include spices (which are a source of odorants) and microorganism growth is not as important as muscle enzymes are for ham flavor development (*1, 15*). These facts cause differences in flavor and in the odorants. In fact, the main qualitative differences between Salami and Parma ham odorants are the presence of allyl methyl sulfide, α-pinene and linalool in the former but not in Parma ham (*6*).

Dry-cured ham is produced by adding salts (sodium chloride, and usually nitrate and nitrite) to raw ham at chilling temperatures, and by ripening (when hams are stable against microorganisms) for several months at higher temperatures; the whole process takes from several months (e.g. Serrano ham) up to more than two years (e.g. Iberian ham). Changes in characteristics of the dry-curing process such as ripening time (*2*) or salt content (*16*) modify sensory characteristics and flavor of hams. Besides processing conditions, there are a number of factors related to raw meat that influence dry-cured ham flavor; among them, the rearing system of pig is critical (*17*) and influences lipid composition and the odorants of hams (*8*). The repercussion of lipid composition on dry-cured ham flavor has been widely researched (reviewed by Gandemer, *18*), but less attention has been paid to the effect of changes in other important flavor precursors such as amino acids. Reactions at high temperature involving important precursors of the meaty-smelling sulfur or nitrogen containing compounds such as cysteine (*19, 20*) or proline (*21*) have been researched and cysteine has been checked to enhance the flavor of wheat extrudates (*22, 23*). Some works focused on cooked meat flavor have reported the effect of adding several precursors to homogenized raw meat before cooking (*24-26*). However, no information about their addition to non-cooked but ripened meats is available.

The aim of this work was to study the influence of salt content, the presence of sodium nitrite, the addition of cysteine or proline, and the reaction time on the most abundant compounds identified previously as dry-cured ham odorants of meat samples.

Materials and Methods

Materials

Pork meat (loin) and subcutaneous fat from one outdoor-reared Iberian pig were purchased from a local slaughterhouse. L-cysteine, L-proline (Sigma-Aldrich), sodium chloride (Scharlau), and sodium nitrite (Scharlau) were used. The reference compounds for volatile compound identification were obtained from Sigma-Aldrich.

Sample Preparation

Meat and fat were trimmed of superficial areas, were separately minced, and then were mixed in a proportion of 90% lean and 10% fat, obtaining a homogeneous meat emulsion. 6 g of the emulsion was placed in a 10 mL tube with a screw teflon-lined cap, and 0.6 g (lower salt content) or 1.2 g (higher salt content) of sodium chloride was added. L-cysteine, L-proline, and sodium nitrite were added as indicated in Table 2.

Table 2. Quantities of precursors added to each sample type.

	$NaNO_2$ (mg/Kg)	cysteine (mmoles/Kg)	proline (mmoles/Kg)
control	-	-	-
cysteine	-	55	-
proline	-	-	55
control + $NaNO_2$	100	-	-
cysteine + $NaNO_2$	100	55	-
proline + $NaNO_2$	100	-	55

After mixing the contents, the tubes were tightly stoppered with Teflon-lined caps and kept at 4°C for 7, 14 or 21 days. All samples were made in duplicate. Other temperatures (25, 90 °C) were also tested. Samples were kept at −80 °C until the volatile compound analysis (less than a month).

Volatile Compound Analysis by Solid-Phase Microextraction-Gas Chromatography-Mass Spectrometry (SPME-GC-MS)

Headspace volatile compounds were collected by using a SPME (Supelco Co., Bellefonte, PA) fiber coated with Carboxen-poly(dimethylsiloxane) (75 μm thickness, 10 mm length). Samples were ground with a scalpel and 2 g were weighed into 5-mL (actual volume) screw-capped vials. Vials were tightly stoppered with hole-caps with a Teflon-rubber disk. For extraction, vials were hold at 30°C in a water bath, and the fiber was inserted into the vials and exposed to headspace for 30 min. Sample order was randomized.

Gas chromatography-mass spectrometry analysis was performed on an HP 5890 series II chromatograph (Hewlett-Packard) coupled to an HP 5973 mass spectrometer (Hewlett-Packard) and equipped with an HP-5 capillary column (50 m x 0.32 mm i.d., film thickness = 1.05 mm, Hewlett-Packard). The injector was maintained at 270 °C. After injection, oven conditions were as follows: 35 °C for 10 min, 7 °C min-1 to 150 °C, 15 °C/min to 250 °C, 250 °C for 10 min. The transfer line to the mass spectrometer was maintained at 280 °C. Mass spectra were generated by electronic impact. A solution of n-alkanes (C5-C18) was analyzed under the same conditions to calculate lineal retention indices (LRI). Compounds were identified by comparing mass spectra and LRI with those of reference compounds analyzed under the same conditions. Only the compounds identified that were previously described as dry-cured ham odorants (6, 7) were taken into account.

Data Analysis

When a compound was identified at least in most samples of a group, an analysis of variance with interaction by the General Linear Model procedure was performed to check the effect of salt content (lower and higher), sodium nitrite (presence, absence), added amino acids (control, cysteine and proline) and reaction time (7, 14 and 21 days). When significant and more than two levels in an effect, multiple comparison by the Tukey test were carried out to compare means. Statistic analyses were performed by means of the SPSS version 11.0.

Results and Discussion

Among the different temperatures tested initially (4, 25 and 95 °C), 4 °C was chosen on the basis of odor characteristics. The higher temperatures (25 and 95 °C) yielded samples without similarity to dry-cured ham odor. The odor of samples at 4 °C showed some of the characteristics of dry-cured ham, although in some cases the buttery note (absent in dry-cured ham) was remarkable. This temperature is similar to the temperature applied during the first stages of the dry-curing process.

The volatile compounds of these samples were analyzed using SPME-GC-MS. 15 of the compounds included in Table 1 were identified in most samples of at least a group and underwent statistical analysis: butane-2,3-dione, acetic acid, 3-methylbutanal, pentanal, hexanal, butanoic acid, hex-(E)-2-enal, 3-methylbutanoic acid, heptan-2-one, hept-(E)-2-enal, octanal, oct-(E)-2-enal, 1-octen-3-ol, and non-(E)-2-enal. The effect of sodium chloride content, sodium nitrite, added amino acids and reaction time on these compounds is shown below. None of the esters, sulfur and nitrogen containing compounds included in Table 1 could be statistically analyzed.

Effect of Sodium Chloride Content

Sodium chloride content is closely related to ham quality, not only because of its direct effect on taste but also because of its effect on flavour development (15). The influence of sodium chloride content on volatile compound generation could be related to its effect on lipid-oxidation reactions, enzyme activity and microbial growth. As expected, salt content affected significantly a number of compounds described previously as dry-cured odorants (Figure 1): seven aldehydes and two acids.

Six straight-chain aldehydes (typical oxidation products) were significantly affected, being more abundant at larger salt content. These data are in agreement with the pro-oxidant activity of sodium chloride in meat systems, which is related to the modification of meat pigments (27) and the disruption of cellular membranes (28).

3-methylbutanal, an amino acid derived compound, was also significantly affected by salt content, but the abundance was larger in the lower salt content group. This fact could be related to the influence of sodium chloride on proteolysis. Sodium chloride affects the activity of muscle enzymes, activity of most enzymes being decreased at larger salt contents (reviewed by Toldrá and Flores, 15).

Effect of Adding Sodium Nitrite

The addition of sodium nitrite caused a general decrease in volatile compounds. The decrease was marked in straight-chain aldehydes, specially on pentanal and hexanal (Figure 2), and could be related to the effective inhibition of lipid oxidation reactions by that nitrite causes in meat (29). The decrease was

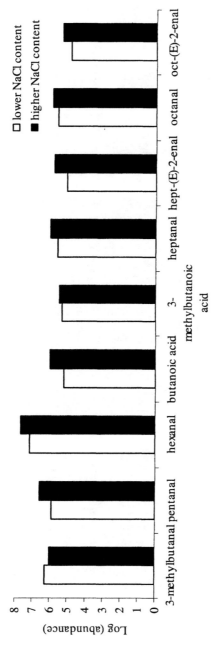

Figure 1. Odorants significantly affected by sodium chloride content at a level p < 0.05.

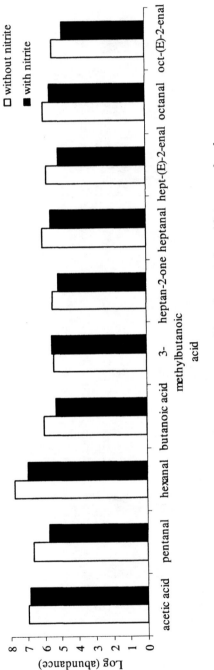

Figure 2. Odorants significantly affected by sodium nitrite addition at a level p < 0.05.

not found for 3-methylbutanal, probably because it is not a lipid-oxidation product and because enzyme activities are not affected by nitrite (*15*).

The effect of sodium nitrite addition on the volatile compounds was barely related to reaction time: no significant interaction appeared, although a slight effect ($p < 0.1$) was found for hexanal and 3-methylbutanoic acid. However, the effect of sodium nitrite was closely related to sodium chloride content: significant interaction ($p < 0.05$) appeared for pentanal, hexanal, heptanal and oct-(E)-2-enal, and a slight effect ($p < 0.1$) for hept-(E)-2-enal and octanal.

Effect of Adding Amino Acids

The addition of cysteine caused a general decrease on most odorants, differences being significant for butane-2,3-dione, 3-methylbutanal and 3-methylbutanoic acid (Figure 3). Therefore, the decrease in odorant generation was specially marked on amino acid derived compounds. Cysteine is an important precursor of sulfur containing compounds (*19*) and it is easily degraded during ham processing: in fact, it is the only amino acid of meat not found as free amino acid in dry-cured ham (*1*). However, none of the odorants from Table 1 typically generated from cysteine was found by using the SPME-GC-MS, probably because they appeared at too low concentration and because of the difficulty of identifying these compounds. In any case, an increase in the meaty note of samples was found when a sensoryanalysis was performed (*30*).

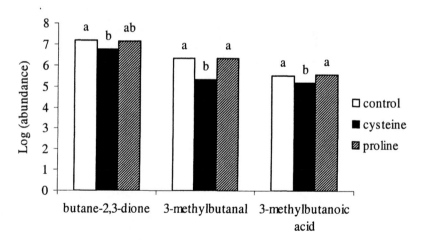

Figure 3. Odorants significantly affected by amino acid addition at a level $p <$ 0.05. Different superscript letter indicated significant differences at a level level $p < 0.05$.

No significant differences were found for any compound when proline was added (Figure 3). Proline is an important precursor of the nitrogen containing compounds included in Table 1 but it was not possible to identify them by using SPME-GC-MS. However, proline addition caused an increase in the cured note when a sensory analysis was performed (*30*).

Effect of Reaction Time

Reaction time influenced the odorant profile. The SPME-GC-MS data revealed significant differences for some compounds previously described as ham odorants (Figure 2): The largest values appeared in the 14-days group, and after that a significant decrease was found. This tendency was also found in other odorants (heptanal, oct-(*E*)-2-enal, butane-2,3-dione) but was not significant.

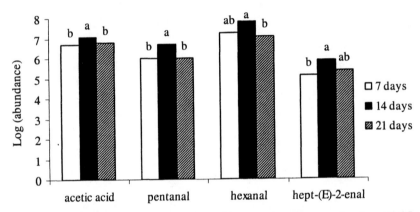

Figure 4. Odorants significantly affected by reaction time at a level p < 0.05.
Different superscript letter indicated significant differences at a level level p <
0.05.

The largest values for the 14-days group matched with the largest scores for the rancid note (*30*). In fact, pentanal and hexanal (two usual indicators of lipid oxidation in meat, *11*) were much more abundant in the 14-days group. It is known that during ripening of hams an increase of lipid oxidation volatile compounds takes place, and it is also followed by a important decrease (*2*). The decrease in some compounds could be related to the decrease of lipid oxidation caused by the depletion of the oxygen available in the tubes and to the lost of odorants by reactions with other components.

References

1. Flores, M.; Spanier, A.M.; Todra, F. In *Flavor of Meat, Meat Products and Seafoods*; Shahidi, F., Ed.; Blackie Academic and Professional: Glasgow, 1998; pp 320-341.
2. Ruiz, J.; Muriel, E.; Ventanas, J. In *Research advances in the quality of meat and meat products*; Toldrá, F., Ed.; Research Signpost: Kerala, India, 2002; pp 289-309.
3. Berdagué, J.L. ; Bonnaud, N. ; Rousset, S. ; Touraille, C. Proceeding 37th International Congress of Meat Science and Technology, Kulmbach, 1991; 1135.
4. Berdagué, J.L. ; Bonnaud, N. ; Rousset, S. ; Touraille, C. *Meat Sci.* **1993**, *34*, 119-129.
5. Flores, M.; Grimm, C.C.; Toldrá, F.; Spanier, A.M. *J. Agric. Food Chem.* **1997**, *45*, 2178-2186.
6. Blank, I.; Devaud, S.; Fay, L.B.; Cerny, C.; Steiner, M.; Zurbriggen, B. In *Aroma active compounds in foods*; Takeoka, G.R.; Güntert, M.; Engel K.H., Eds.; Symposium series 794; ACS: Washington, 2001; pp 9-20.
7. Carrapiso, A.I.; Ventanas, J.; García, C. *J. Agric. Food Chem.* **2002**, 50, 1996-2000.
8. Carrapiso, A.I.; Jurado, A.; Timón, M.L.; García, C. *J. Agric. Food Chem.* **2002**, 50, 6453-6458.
9. Gasser, U.; Grosch, W. *Z. Lebensmit. Untersuch. Forsch.* **1988**, *186*, 489-494.
10. Bredie, W.L.P.; Ammann, C.R.; Bult, J.H.F. In *Frontiers of flavor science*; Schieberle, P.; Engel, K.H., Eds.; WB-Druck Press GmbH: Rieden am Forggensee, Germany, 2000; pp 220-224.
11. Shahidi, F. In *Headspace Analysis of Foods and Flavors. Theory and Practice*; Rouseff, R.L.; Cadwallader, K.R., Eds.; Kluwer academic: New York, 2001, pp 113-123.
12. Carrapiso, A.I.; García, C. Extremadura University. Unpublished.
13. García, C.; Martín, A.; Timón, M.L.; Córdoba, J.J. *Letters Appl. Microbiol.* **2000**, *30*, 61-66.
14. Kerscher, R.; Grosch, W. *Z. Lebensmit. Untersuch. Forsch.* **1997**, *204*, 3-6.
15. Toldrá, F. ; Flores, M. *Crit. Rev. Food Sci.* **1998**, *38*, 331-352.
16. Arnau, J.; Guerrero, L. ; Gou, P. *J. Sci. Food Agric.* **1997**, *74*, 193-198.
17. Cava, R.; Ventanas, J.; Ruiz, J.; Andrés, A.I.; Antequera, T. *Food Sci. Technol. Int.* **2000**, *6*, 235-242.
18. Gandemer, G. *Meat Sci.* **2002**, *62*, 309-321.
19. Hofmann, T.; Schieberle, P. *J. Agric. Food Chem.* **1995**, *43*, 2187-2194.
20. Cerny, C.; Davidek, T. *J. Agric. Food Chem.* **2003**, *51*, 2714-2721.

21. Blank, I.; Devaud, S.; Matthey-Doret, W.; Robert, F. *J. Agric. Food Chem.* **2003**, *51*, 3643-3650.
22. Bredie, W.L.P.; Hassel, G.M.; Guy, R.C.E.; Mottram, D.S. *J. Cereal Sci.* **1997**, *25*, 57-63.
23. Hwang, C.F.; Riha III, W.E.; Jin, B.; Karwe, M.V.; Hartman, T.G.; Daun, H.; Ho, C.H. *Lebensm.Wiss. Technol.* **1997**, *30*, 411-416.
24. Mottram, D.S.; Madruga, M.S. In *Trends in flavour research*; Maarse, H.; van der Heij, D.G., Eds.; Elsevier: Amsterdam, 1994; pp 339-344.
25. Farmer, L.J.; Hagan, T.D.H.; Paraskevas, O. In *Flavour science. Recent development*; Taylor, A.J.; Mottram, D.S., Eds; Royal Society of Chemistry: Cambridge, 1996, pp 225-230.
26. Farmer, L.J.; Hagan, T.D.J. Proceeding 48th International Congress of Meat Science and Technology, Rome, 2002; 322-323.
27. Kanner, J.; Salan, M.A.; Harel, S.; Shegalovich, I. *J. Agric. Food Chem.* **1991**, *39*, 242-246.
28. Shomer, I.; Weinberg, J.G.; Vasilever, R. *Food Microstr.* **1987**, *6*, 199-207.
29. Ramarathnam, R.; Rubin, L.; Diosady, L.L. *J. Agric. Food Chem.* **1991**, *39*, 344-350.
30. Carrapiso, M.R.; Martín, L.; Carrapiso, A.I. Extremadura University. Unpublished.
31. Rychlik, M.; Schieberle, P.; Grosch, W. *Compilation of odor thresholds, odor qualities and retention indices of key food odorants*. Deutsche Forschungsanstalt für Lebensmittelchemie: Garching, Germany, 1998.

Chapter 6

Aroma and Amino Acid Composition of Hydrolyzed Vegetable Protein from Rice Bran

Arporn Jarunrattanasri[1], Keith R. Cadwallader[2,*], and Chockchai Theerakulkait[3]

[1]Department of Agro-Industry, Faculty of Agriculture Natural Resources and Environment, Naresuan University, Pitsanuloke 65000, Thailand
[2]Department of Food Science and Human Nutrition, University of Illinois, 1302 West Pennsylvania Avenue, Urbana, IL 61801
[3]Department of Food Science, Kasetsart University, Jatujak, Bangkok 10900, Thailand

Hydrolyzed vegetable protein (HVP) is an important ingredient of many process flavors. HVP was produced from rice bran protein by acid hydrolysis with aqueous 4.0 N HCl for 72 h at 95 °C. Predominant free amino acids were glutamic acid (16.7% of total), arginine (11.8%), aspartic acid (10.1%), alanine (8.1%), leucine (7.6%) and glycine (6.8%). Aroma components were identified by GC-olfractometry (GCO), including aroma extraction dilution analysis (AEDA) and GCO of decreasing dynamic headspace samples (GCO-DHS), and by GC-MS. The overall "malty", "soy sauce-like" and "smoky" aroma character of rice-bran HVP was dominated by Strecker aldehydes and acidic/phenolic compounds. Methylpropanal and 2/3-methylbutanal were responsible for malty and dark chocolate-like notes, while 3-(methylthio)propanal provided a cooked potato-like character and phenylacetaldehyde imparted a rosy, styrene-like note. Acidic compounds with high odor impact included sotolon, and phenylacetic and 3-methylbutanoic acids. 2-Methoxyphenol, and to a lesser extent 2,6-dimethoxyphenol, provided intense smoky notes. 2-Acetyl-1-pyrroline also exhibited a relatively high odor impact in RB-HVP.

Introduction

Hydrolyzed vegetable protein (HVP) is one of the earliest known forms of thermal reaction or process flavors (*1,2*). HVPs can been produced by acid (HCl) o r e nzyme (proteolytic) h ydrolysis o f a p rotein s ource (usually o f p lant origin) t o f orm principally amino acids (*1,3-5*), which, themselves, can impart taste (e.g. monosodium glutamate) or participate in subsequent thermal reactions, e.g. Maillard reaction, to form aroma compounds (*6,7*). Among the numerous process parameters involved in the production of HVP, the substrate or protein source material may have a great impact on the resulting amino acid profile and flavor characteristics of the final product (*1,3*).

Rice bran is the outer brown layer, including the rice germ, removed during milling of brown rice to produce white rice. Rice bran is rich i n h igh q uality protein, B vitamins, and oil (*8*). Rice bran represents an underutilized and potentially inexpensive and plentiful source of plant p rotein f or p roduction o f HVPs having unique amino acid profiles and flavor characteristics.

Gas chromatography-olfactometry (GCO) has been used extensively for the identification of characteristic aroma components of foods (*9,10*). Aroma extract dilution analysis (AEDA) is a GCO technique in which serial dilutions (e.g. 1:3) of an aroma extract are evaluated by GCO. In AEDA, the highest dilution at which an odorant is last detected during GCO, so-called flavor dilution (FD) factor, is used as a measure of its odor potency (*9*). One potential drawback to AEDA is that the technique is limited to the analysis of components of intermediate and low volatility. To overcome this limitation, AEDA results have been complemented by results of GCO of decreasing dynamic headspace (DHS) and decreasing static headspace (GCO-H) samples (*10,11*)

The objective of the present study was to determine the key aroma components and amino acid composition of an HVP produced by acid hydrolysis of rice bran protein concentrate.

Experimental Procedures

Chemicals

Analytical-grade reference compounds and solvents were obtained commercially (Aldrich Chemical Co., St. Louis, MO), except (*E*)-β-Damascenone was provided by Firmenich Co. (Princeton, N J) a nd 2 -acetyl-1-pyrroline was a gift from Dr. R. Buttery (USDA, ARS, WRRC, Albany, CA). Dimethyltetrasulfide was synthesized using a published procedure (*12*). Reagents used for amino acid analyses w ere o btained f rom B eckman C oulter, Inc. (Fullerton, CA).

Defatted rice bran (DRB)

Rice bran was obtained from a local rice processor in Bangkok, Thailand. It was defatted with hexane to produce DRB using the procedure described by Wang et al. (*13*), then air dried, ground in a sample mill and sieved through 80 mesh screen. DRB was packed in polyethylene bags and stored at 5°C.

Rice bran protein concentrate (RBPC)

RBPC was prepared from DRB by alkali extraction and isoelectric precipitation using the procedure described by Gnanasambandam and Hettiarachchy (*14*).

Preparation of Rice Bran Hydrolyzed Vegetable Protein (RB-HVP)

RB-HVP was prepared using a modification of published procedures (*15,16*). RBPC (200 g, containing approximately 50 % (w/w) protein) plus 500 mL of aqueous 4.0 N HCl were placed in a glass 2-L amber bottle. The bottle was flushed with N_2 (~100 mL/min) for 10 min and sealed with a Teflon-lined cap. Hydrolysis was conducted at 95°C for 72 h. The pH of the hydrolysate was adjusted to 6.0 with aqueous 10.0 N NaOH and filtered through celite (0.1 % w/v). The hydrolysate was spray-dried (inlet temperature 200-210 °C, outlet temperature 110-120 °C) using a Niro Atomizer spray drier. RB-HVP was stored at −70°C until analysis. It should be noted that no attempt was made to 'age' the HVP prior to spray drying, since a previous study showed there to be no apparent advantage from a flavor standpoint (*5*).

Determination of Total and Free Amino Acids

For determination of total amino acids RBPC was first hydrolyzed with 10-mL of aqueous 6.0 N HCl at 105°C for 22 h, then neutralized and filtered.

Free amino acids were analyzed by first dissolving samples (0.05-0.1 g/mL) in 0.2 N citrate buffer, pH 2.2, and the mixture shaken for 15 min at room temperature. Mixture was centrifuged at 7000 x g for 5 minutes and supernatant deproteinized with 3.5 % sulfosalicylic acid solution by mixing at a 1:1 ratio. Solution was chilled and shaken to maximize precipitation of unwanted protein. After thawing, the solution was centrifuged at 7000 x g for 5 minutes.

Filtrate (total amino acids) or supernatant (free amino acids) from above was diluted (20:1) with 0.2 N lithium citrate buffer, pH 2.8, prior to analysis using a Beckman 6380 amino acid analyzer (Beckman Coulter, Inc.) equipped with a 10 cm ion exchange column and lithium citrate buffer supplied by Beckman. Detection was via post column derivatization using ninhydrin. Calibration was via mixed external free amino acid standards.

Isolation of Volatiles

RB-HVP was subjected to direct solvent extraction followed by high vacuum distillation (17). Ten g of PB-HVP was hydrated with 100 mL of odor-free distilled-deionized water. The mixture was spiked with 25 μL of an internal standard solution (5.0 μg/μL of 2-ethyl butanoic acid; 3.3 μg/μL of 4-*tert*-amyphenol and 4.3 μg/μL of 2-methyl-3-heptanone in methanol). The solution was extracted with ether (2 x 50 mL). Extractions were conducted for duplicate samples.

Compound class fractionation. E ach e ther e xtract w as s ubjected to high vacuum distillation as previous described (17). The volatile extract (distillate) was concentrated to 20 mL under a gentle stream of nitrogen gas (N_2). It was then washed with aqueous sodium carbonate solution (Na_2CO_3) (5 %w/v; 3 x 20 mL). The upper ether phase, containing the neutral/basic volatiles, was collected and concentrated to 10 mL under N_2. It was then dried over anhydrous Na_2SO_4 (2 g) and concentrated to 0.2mL under N_2. The pooled aqueous phase (bottom layer) was acidified with aqueous HCl (10% w/v) to pH 3 and the acidic volatiles extracted with ether (3 x 15 mL). The pooled ether extract was then concentrated to 10 mL under N_2, dried over anhydrous Na_2SO_4, and then further concentrated to 0.2 mL under N_2.

Aroma Extract Dilution Analysis (AEDA)

The GCO system consisted of a HP6890 GC (Agilent Technologies, Palo Alto, CA) equipped with a flame ionization detector (FID), a sniffing port (DATU, G eneva, N Y) a nd c ool on-column injector. Each extract (2 μL) was injected into a capillary column (Stabilwax™ DA, 15 m length x 0.32 mm i.d. x 0.5 μm film thickness (d_f), Restek Corp., Bellefonte, PA; or DB-5MS 15 m length x 0.32 mm i.d. x 0.5 μm film; J & W Scientific, Folson, CA). GC oven temperature was programmed from 35 °C to 200 °C at a rate of 10 °C/min with initial and final hold times of 5 and 30 min, respectively. Carrier gas was helium at a constant flow of 2.2 mL/min. Two experienced panelists conducted GCO. The extracts containing the neutral/basic and acidic volatiles were diluted stepwise with diethyl ether at a ratio of 1:3 (v/v). The aroma extract dilution procedure was performed until no odorants were detected by GCO. The highest dilution was defined as flavor dilution (FD) factor (9).

Gas Chromatography-Olfactometry of Decreasing Dynamic Headspace Samples (GCO-DHS)

GCO-DHS was conducted on a n H P6890 G C (Agilent T echnologies, I nc.) equipped with a Tekmar 3000 purge and trap concentrator/cryofocusing module

(Tekar Co., Cincinnati, OH), flame ionization detector, and sniff port. RB-HVP (0.1 g) dissolved in 5 mL of odorless-distilled was placed in a 25 mL headspace sampling tube. The sample was preheated to 60°C for 5 min. Volatiles were purged onto a Tenax TA trap (Tekmar Co.) at 60°C with helium (40 mL/min) for 2.5, 7.5, 22.5 and 67.5 min, which corresponded to purge volumes of 100, 300, 900 and 2700 mL, respectively. All other purge and trap conditions were the same as previously described by Wu and Cadwallader (18). G CO conditions were same as described above for AEDA. A freshly prepared sample was used for each analysis.

Gas Chromatography-Mass Spectrometry

The system consisted of an HP6890 GC/5973 mass selective detector (MSD, Agilent Technologies, Inc.). Separations were performed on a fused silica capillary column (Stabiliwax DA, 30 m length x 0.25 mm i.d. x 0.5 μm film, Restek Corp.; or HP-5MS, 30 m length x 0.25 mm i.d. x 0.5 μm film, Agilent Technologies, Inc.). Carrier gas was helium at a constant flow of 1 mL/min. The oven temperature was programmed from 35°C to 200°C at a rate of 3 °C/min with initial and final hold times of 5 and 45 min, respectively. MSD conditions were as follows: capillary direct interface temperature, 280°C; ionization energy, 70 eV; mass range, 33-350 a.m.u; EM voltage (Atune+200 V); scan rate, 2.2 scans/s. Each extract (2μL) was injected in the cool on-column mode.

The DHS GC-MS system was same as above except that injections were performed by the Tekmar 3000 system. For DHS, all conditions were same as earlier described except that the RB-HVP sample (0.1 g) was spiked with 10 μL of internal standard solution (3.4 ng/μL of 2-methyl-3-heptanone in methanol) prior to analysis and prepurge and purge times were 2 and 10 min, respectively.

Identification of Odorants

Positive compound identifications were made by comparing retention indices (RI), mass spectra, and odor properties of unknowns with those of authentic standard compounds analyzed under identical conditions. Tentative identifications were based on matching the RI values and odor properties of unknowns against those of authentic standards. Retention indices were calculated by using a series of n-alkanes (19).

Quantification of Odorants

Concentrations were calculated based on relative GC-MS abundance of internal standards, 2-methyl-3-heptanone for neutral/basic compounds and 2-

ethyl butanoic acid for acidic compounds, except 2-methoxyphenol and 2,6-dimethoxyphenol standardized against 4-*tert*-amyphenol.

Results and Discussion

Amino acid composition

Total and free amino acid compositions for RBPC and RB-HVP, respectively, are shown in Table I. RB-HVP contained about 38% (w/w, dry weight basis) free amino acids. Based on the total amino acids available in RBPC, about 68% of the available free amino acids were recovered in the spray-dried RB-HVP. Six amino acids, including glutamic acid, arginine, aspartic acid, alanine, leucine and glycine were in highest abundance and accounted for over 61% of the total free amino acids in RB-HVP. In particular, the flavor enhancer glutamic acid was in highest abundance (16.7% of total free amino acids). Glutamic acid is often found in highest abundance in HVPs. For example, glutamate may comprise as much as 22-25% of the free amino acids of soy HVP (*3,20*), about 33% of wheat HVP (*3*), and about 14% of HVP derived from alfalfa flour (*20*). In addition to their taste impact, free amino acids also serve as important precursors in the thermal generation of an extensive range of characteristic meaty and roasty aroma compounds associated with cooked meat flavor.

Aroma components of RB-HVP

Odorants of intermediate and low volatility were isolated from the spray dried RB-HVP by direct solvent extraction, followed by a high-vacuum distillation cleanup step to remove any nonvolatile residue from the aroma extract. Twenty-eight odorants with were detected by GCO (Table II). Predominant neutral/basic odorants (average log$_3$FD-factors $\geq$ 3.0) were 3-(methylthio)propanal (**8**, *potato*), 2-acetyl-1-pyrroline (**6**, *roasty/popcorn*), phenylacetaldehyde (**11**, *rosy/honey*), o-aminoacetophenone (**16**, *corn tortilla/grape*) and dimethyltetrasulfide (**10**, *sulfurous/garlic salt*). Acidic odorants detected at average log$_3$FD-factors $\geq$ 3.0 were 2/3-methylbutanoic acid (**19/20**, *sweaty/dried fruit*), Furaneol® (**24**, *burnt sugar/caramel*), sotolon (**25**, *curry*), 2,6-dimethoxyphenol (**26**, *smoky*) and phenylacetic acid (**27**, *rosy/plastic*). Other important odorants included 3-methylbutanal (**2**, *malty/dark*

**Table I. Amino Acid Composition of Rice Bran Protein Concentrate
(RBPC) and Rice Bran Hydrolyzed Vegetable Protein (RB-HVP)**

| | *percent (dry wt basis)[a]* | | *percent of total* | |
| | RBPC (total amino acids) | RB-HVP (free amino acids) | RBPC (total amino acids) | RB-HVP (free amino acids) |
amino acid				
aspartic acid	4.87	3.86	8.72	10.12
threonine	2.27	1.64	4.06	4.30
serine	3.19	2.26	5.71	5.92
glutamic acid	8.84	6.38	15.83	16.72
proline	2.59	1.84	4.64	4.82
glycine	3.53	2.58	6.32	6.76
alanine	4.10	3.09	7.34	8.10
valine	3.13	1.75	5.60	4.59
methionine	1.11	0.83	1.99	2.18
isoleucine	1.95	1.05	3.49	2.75
leucine	4.70	2.91	8.42	7.63
tyrosine	2.34	0.61	4.19	1.60
phenylalanine	2.56	1.55	4.58	4.06
histidine	1.93	1.08	3.46	2.83
lysine	2.83	2.22	5.07	5.82
arginine	5.91	4.50	10.58	11.80
total	55.85	38.15	100.0	100.0

[a]Average (n = 2)

chocolate), dimethyltrisulfide (**7**, *sulfurous/cabbage*), 2-acetyl-2-thiazoline (**13**, *popcorn*), butanoic acid (**18**, *fecal/cheesy*), hexanoic acid (**22**, *sweaty*), and 2-methoxyphenol (**23**, *smoky*).

Highly volatile compounds that may be overlooked by AEDA because of losses during extraction and sample workup were evaluated by GCO-DHS. Results indicated methanethiol (**29**, *sulfurous/rotten*), dimethylsulfide (**30**, *sulfurous/fresh corn*) and 2,3-butanedione (**31**, *buttery*) as additional predominant headspace odorants and confirmed the importance of compounds **1-3, 5-8, 11, 17** and **23** (Table III) in the aroma of RB-HVP.

Twenty-one volatile compounds were selected for quantitation (Table IV). Two procedures were employed depending upon the volatilities of the chosen analytes. DHS-GC-MS was used for highly volatile compounds **1-3** and **29-31**;

Table II. Predominant Odorants Detected by Aroma Extract Dilution Analysis of Rice Bran Hydrolyzed Vegetable Protein (RB-HVP)

no.[a]	compound	odor description[b]	retention index[c] FFAP	retention index[c] DB-5MS	average $\log_3$ FD-factor[d]
	Neutral/Basic				
1	methylpropanal	malty, dark chocolate	824	898	1.0
2	3-methylbutanal	malty, dark chocolate	912	645	2.0
3	2-methylbutanal	malty, musty, chocolate	927	659	<1.0
4	hexanal	green, cut-grass	1086	801	1.0
5	1-octen-3-one[e]	Mushroom	1301	979	1.5
6	2-acetyl-1-pyrroline	roasty, popcorn	1335	921	4.0
7	dimethyltrisulfide	sulfurous, cabbage	1367	972	2.5
8	3-(methylthio)propanal	Potato	1456	909	7.0
9	(E)-2-nonenal[e]	Cucumber	1528	1161	1.0
10	dimethyltetrasulfide	sulfurous, garlic salt	1634	1216	3
11	phenylacetaldehyde	rosy, honey	1648	1046	3.0
12	β-damascenone[e]	Applesauce	1821	1387	1.0
13	2-acetyl-2-thiazoline	Popcorn	1766	1120	2.0
14	2-phenylethanol	rosy, wine	1913	1116	1.0
15	benzothiazole	Rubbery	1958	1232	1.5
16	o-aminoacetophenone	corn tortilla, grape	2202	1307	3.0
	Acidic				
17	acetic acid	vinegar	1443	1290	1.5
18	butanoic acid	fecal, cheesy	1622	878	2.0

No.	Compound	Odor quality[b]	RI[c]	RI[c]	FD[d]
19/20	2/3-methylbutanoic acid	sweaty, dried fruit	1660	811	3.0
21	pentanoic acid	Swiss cheese, sweaty	1731	869	1.5
22	hexanoic acid	sweaty	1839	1021	2.5
23	2-methoxyphenol	smoky	1867	1092	2.5
24	Furaneol®[f]	burnt sugar, caramel	2022	1214	4.0
25	sotolon[e,g]	curry	2198	1135	5.0
26	2,6-dimethoxyphenol	smoky	2272	1357	5.0
27	phenylacetic acid	rosy, plastic	2556	1292	5.0
28	vanillin[h]	sweet, vanilla	2577	1370	1.5

[a]Numbers correspond to those in Tables III and IV. [b]Odor quality as perceived during GCO. [c]Retention indices calculated from GCO data, FFAP = Stabilwax™ DA column. [d]Average log₃ flavor dilution (FD) factor (n = 2), neutral/basic fraction determined on DB-5MS and acidic fraction determined on Stabilwax™ DA column. [e]Compound tentatively identified based on comparison of odor property and retention indices with reference compound. [f]2,5-dimethyl-4-hydroxy-3(2H)-furanone. [g]3-hydroxy-4,5-dimethyl-2(5H)-furanone. [h]3-methoxy-4-hydroxybenzaldehyde

Table III. Predominant Odorants Detected by Decreasing Dynamic Headspace Volumes of Rice Bran Hydrolyzed Vegetable Protein (RB-HVP)

| no.[a] | compound | odor description[b] | Retention Index[c] | | average $\log_3$ FD-factor[d] |
			FFAP	DB-5MS	
29	methanethiol	sulfurous, rotten	650	< 600	2
30	dimethylsulfide	sulfurous, fresh corn	755	< 600	2
1	methylpropanal	malty, dark chocolate	824	898	1.0
2	2-methylbutanal	malty, musty, chocolate	913	663	1.5
3	3-methylbutanal	malty, dark chocolate	931	642	2
31	2,3-butanedione	buttery	913	663	1.5
5	1-octen-3-one[e]	mushroom	1309	980	<1
6	2-acetyl-1-pyrroline	roasty, popcorn	1340	--	1
7	dimethyltrisulfide	sulfurous, cabbage	1375	971	2
17	acetic acid	vinegar	1450	--	<1
8	3-(methylthio)propanal	potato	1454	905	1.5
11	phenylacetaldehyde	rosy, honey	1650	1044	<1
23	2-methoxyphenol	smoky	1866	1091	2.5

[a]Numbers correspond to those in Tables II and IV. [b]Odor quality as perceived during GCO. [c]Retention indices calculated from GCO data, FFAP = Stabilwax™ DA column. [d]Average $\log_3$ flavor dilution (FD) factor (n = 2) determined on Sabilwax DA column; An FD factor corresponded to the highest purge gas volume tested (2700 mL) divided by lowest purge gas volume (900, 300 or 100 mL) in which compound was last detected by GCO. [e]Compound tentatively identified based on comparison of odor property and retention indices with reference compound.

while the remaining compounds in Table IV were determined by solvent extraction-GC-MS. The concentration values in Table IV are relative values based on recovery of the internal standards. Some odorants (i.e., **5**, **9**, **10**, **12**, **13**, **14**, **16**, **24**, and **25**) were excluded from Table IV because they were below the detection limits of the quantification methods.

Overall, the calculated OAVs correlated well with the GCO findings, with the notable exception of acidic compounds **24** and **25**, which had much lower OAVs than expected based on their high average $\log_3 FD$ factors. The acidic compounds **17**, **20** and **25** were in highest abundance, but may not contribute much to the overall aroma of RB-HVP due to their low OAVs. Based on their relatively high OAVs, butanoic acid (**12**) and 2-methoxyphenol (**21**) may be the only acidic compounds that contribute greatly to the aroma of RB-HVP. However, it is possible that Furaneol® (**24**) and sotolon (**25**), which were not quantified, may contribute to RB-HVP aroma. Especially compound (**25**), which has an extremely low odor detection threshold of 0.001 ng/mL (*21*). Among the neutral/basic odorants, dimethylsulfide (**30**), 2- (**3**) and 3-methylbutanal (**2**), dimethyltrisulfide (**7**) and 3-(methylthio)propanal (**8**) had very high OAVs, and may make the greatest contribution to the aroma of RB-HVP. In addition, compounds **1**, **6**, **11**, **29** and **31** with their relatively high OAVs also may contribute to the overall aroma of RB-HVP.

Results of GCO and the calculated OAVs indicate that compounds having a wide range of volatilies contribute to the aroma of RB-HVP. The Maillard reaction, specifically the Strecker degradation of free amino acids to form volatile aldehydes, plays a key role in the generation of the aroma of RB-HVP. The overall "meaty", "malty" and "soy sauce-like" character of RB-HVP can be mostly attributed to the Strecker aldehydes. The predominance of Strecker aldedhyes, such as methylpropanal (from valine), 2- and 3-methylbutanal (from isoleucine and leucine, respectively), 3-(methylthio)propanal (from methionine) and phenylacetaldehyde (from phenylalanine), are in general agreement with published literature on the flavor chemistry of HVP (*4,22*). Especially important may be the Strecker degradation of methionine to form 3-(methylthio)propanal, which may further degrade by oxidative means to yield methanethiol, dimethylsulfide, dimethyltrisulfide and dimethytetrasulfide (*23, 24*). On the other hand, the "smoky" aroma character of rice-bran HVP may be due degradation of phenolic compounds (e.g. lignin) to form principally 2-methoxyphenol and 2,6-dimethoxyphenol. In a previous study, the former compound was considered to be an off flavor of soy HVP (*22*). It may be possible to produce RB-HVP having less pronounced "smoky" flavor if activated charcoal treatment of the hydrolysate is conducted prior to spray-drying. In addition to the above compounds, 2-Acetyl-1-pyrroline, a Maillard reaction product from proline, also may contribute to the aroma of RB-HVP.

Table IV. Concentrations and Odor Activity Values of Selected Volatile Components of Rice Bran Hydrolyzed Vegetable Protein (RB-HVP)

no.[a]	compound	threshold (ng/g in water)	solvent extraction[b]		DHS[c]	
			Concn (ng/g)	OAV[d]	Concn (ng/g)	OAV[e]
29	Methanethiol	0.2[f]	-[e]	n.a.	0.32	1.6
30	dimethylsulfide	0.3[g]	--	n.a.	4.2	14
31	2,3-butanedione	2.3[h]	--	n.a.	230	100
1	methylpropanal	1[i]	--	n.a.	44	44
2	3-methylbutanal	0.35[j]	730	2090	2020	5770
3	2-methylbutanal	1[i]	140	140	800	800
4	hexanal	4.5[j]	14	3	--	n.a.
6	2-acetyl-1-pyrroline	0.1[k]	7	70	--	n.a.
7	dimethyltrisulfide	0.01[l]	9	900	--	n.a.
8	3-(methylthio)propanal	0.2[j]	650	3300	--	n.a.
11	phenylacetaldehyde	4[j]	320	80	--	n.a.
15	benzothiazole	80[k]	240	3	--	n.a.
17	acetic acid	50000[m]	1310	<0.1	--	n.a.
18	butanoic acid	50[m]	608	12	--	n.a.
19	3-methylbutanoic acid	250[g]	530	2	--	n.a.
21	pentanoic acid	1207[n]	540	<0.1	--	n.a.
22	hexanoic acid	3000[g]	3950	0.5	--	n.a.
23	2-methoxyphenol	3[k]	210	70	--	n.a.

26	2,6-dimethoxyphenol	1850[o]	760	<1	- -	n.a.
27	phenylacetic acid	10000[p]	1320	0.4	- -	n.a.
28	vanillin[h]	30[n]	60	2	- -	n.a.

[a]Numbers correspond to those in Tables II and III. [b]Relative concentration (average, n = 2) by solvent extraction-high vacuum distillation. [c]Relative concentration (average, n = 2) by dynamic headspace sampling method. [d]Odor activity value (OAV) was calculated by dividing compound concentration by its published odor detection threshold. [e]Concentration not determined. [f]Guth and Grosch (25). [g]Buttery (26). [h]Fors (27). [i]Amoore et al. (28). [j]Guadagni et al. (29). [k]Buttery et al. (30). [l]Buttery et al. (31). [m]Larsen and Poll (32). [n]Karagül-Yüceer et al. (33). [o]Wasserman (34). [p]Maga (35).

This compound may be uniquely high in RB-HVP since it is a known character-impact compound of aromatic rice (*30*).

Conclusions

Rice bran protein concentrate can be effectively acid hydrolyzed to produce RB-HVP having a unique free amino acid profile, with a slightly reduced monosodium glutamate balance compared to soy and wheat HVPs. The majority of the character-impact components of RB-HVP were derived via Strecker degradation of the available free amino acids, with some potential off-notes being caused by presence of 2-methoxyphenol and 2,6-dimethoxyphenol. In order to fully understand the potential of RB-HVP, it is recommended that the product be further characterized by sensory analysis and evaluated in an actual process flavor application.

Literature Cited

1. Manley, C. Process flavors. In *Source Book of Flavors*, 2nd ed. Reineccius, G., Ed., Chapman & Hall: New York, 1994; pp. 139-154.
2. Manley, C.H. and Ahmedi, S. *Trends Food Sci. & Technol.*, **1995**, *6*, 46-51.
3. Swaine Jr., R.L. *Perfumer & Flavorist* **1993**, *18*, 35-38.
4. Aaslyng, M.D.; Martens, M.; Poll, L.; Nielson, P.M.; Flyge, H.; Larsen, L.M. *J. Agric. Food Chem.* **1998**, *46*, 481-489.
5. Aaslyng, M.D.; Larsen, L.M.; Nielsen, P.M. *Z. Lebemsm. Unters. Forsch A* **1999**, *208*, 355-361.
6. Mottram, D.-S. In *Thermally Generated Flavors, Maillard, Microwave and Extrusion Processes*; Parliment, T.H., Morello, M.J., McGorrin, R.J., Eds.; American Chemical Society: Washington, DC, 1994; pp. 104-125.
7. Weenen, H. *Food Chem.* **1998**, *62*, 393-401.
8. Juliano, B.O. In *Rice: Chemistry and* Technology; Juliano, B.O., Ed.; American Association of Cereal Chemists, Inc.: St. Paul, MN, 1985; pp. 647-687.
9. Grosch, W. *Trends Food Sci. Technol.* **1993**, *4*, 68-71.
10. Blank, I. *In Flavor, Fragrance and Odor Analysis*; Marsili, R., Ed.; Marcel Dekker, Inc.: New York, 2002; pp. 297-331.
11. Cadwallader, K.R. and Baek, H.H. In *Food Flavors: Formation, Analysis and Packaging Influences*; Contis, E.T.; Ho, C.T.; Mussinan, C.J.; Parliment, T.H.; Shahidi, F.; Spanier, A.M., Eds.; Elsevier: Amsterdam, 1998; pp. 271-278.
12. Milligan, B.; Saville, B.; Swan, J.M. *J. Chem. Soc.* **1963**, 3608-3614.
13. Wang, M.; Hettiarachchy, N.S.; Qi, M.; Burk, W.; Siebenmorgen, T. *J. Agric. Food Chem.* **1999**, *47*, 411-416.

14. Gnanasambandam, R.; Hettiarachchy, N.S. *J. Food Sci.* **1995**, *60*, 1066-1074

15. Pharm, C.B.; Del Rosario, R.R. *J. Food Technol.* **1983**, *18*, 21-34.

16. Wattanapat, R.; Nakayama, T.; Beuchat, L.R. *J. Food Sci.* **1995**, *60*, 443-445.

17. Karagül-Yüceer, Y.; Drake, M. A.; Cadwallader, K. R. *J. Agric. Food Chem.* **2001**, *49*, 2948-2953.

18. Wu, Y.-F.G.; Cadwallader, K.R. *J. Agric. Food Chem.* **2002**, *50*, 2900-2907.

19. Van den Dool, H; Kratz, P. D. *J. Chromatogr.* **1963**, *11*, 463-471.

20. Džanić, H.; Mujić, I.; Sudarski-Hack, V. *J. Agric. Food Chem.* **1985**, *33*, 683-685.

21. Kobayashi, A. In *Flavor Chemistry-Trends and Developments*; Teranishi, R., Buttery, R.G., Shahidi F., Eds.; American Chemical Society: Washington, DC, 1989; pp. 49-59.

22. Aaslyng, M.D.; Elmore, J.S.; Mottram, D.S. *J. Agric. Food Chem.* **1998**, *46*, 5225-5231.

23. Forss, D.A. *J. Dairy Res.* **1979**, *46*, 691-706.

24. Schutte, L. *CRC Crit. Rev. Food Technol.* **1974**, *4*, 457-505.

25. Guth, H.; Grosch, W. *J. Agric. Food Chem.* **1994**, *42*, 2862-2866.

26. Buttery, R.G. In *Flavor Science – Sensible Principles and Techniques in Flavor Science*; Acree, T., Teranishi, R., Eds.; American Chemical Society: Washington, DC, 1993; pp. 259-286.

27. Fors, S. In *Maillard Reaction in Food Nutrition*; Waller, G.R. and Feather, M.S., Eds.; American Chemical Society: Washington, DC; 1983; pp. 185-286.

28. Amoore, J.E.; Forrester, L.J.; Pelosi, P. *Chem. Senses Flavor* **1976**, *1*, 17-25.

29. Guadagni, G.G.; Buttery, R.G.; Turnbaugh, J.G. *J. Sci. Food Agric.* **1972**, *23*, 1435-1444.

30. Buttery, R.G.; Turnbaugh, J.G. and L. C. Ling, J. *Agric. Food Chem.* **1988**, *36*, 1006-1009

31. Buttery, R.G.; Guadagni, D.G.; Ling, L.C., Seifert, R.M., Lipton, W. *J. Agric. Food Chem.* **1976**, *24*, 829-832.

32. Larsen, M.; Poll, L. *Z. Lebensm. Unters. Forsch.* **1992**, *195*, 120-123.

33. Karagül-Yüceer, Y.; Vlahovich, K.N.; Drake, M.A.; Cadwallader, K.R. J. Agric. Food Chem. **2003**, *51*, 6797-6801.

34. Wasserman, A.E. Organoleptic evaluation of three phenols present in wood smoke. *J. Food Sci.* **1966**, *31*, 1005-1010.

35. Maga, J. A. Taste threshold values for phenolic acids which can influence flavor properties of certain flours, grains and oilseeds. *Cereal Sci. Today* **1973**, *18*, 326-330.

Chapter 7

Reaction Flavors: The Next Generation

Brian Byrne

Natural Advantage, 1050 Cypress Creek Road, Freehold, NJ 07728
(brian.byrne@natural-advantage.net)

Reaction flavors, also known as process flavors, have traditionally been the reaction of heating a protein source and a sugar to produce a mixture of chemicals containing flavor value. The products are the result of complex reactions involving Maillard, Strecker and caramelization reactions, followed by cross-reactions of the initial products.

More specific flavors have been obtained by substituting specific amino acids and specific sugars. An example is the microwave reaction of Glucose, Phenylalanine and Leucine by Byrne and Buckholz[1], to get a chocolate tasting flavor.

In addition to obtaining flavorful mixtures, some chemicals can be obtained in such good yields that they are of synthetic utility. Natural Isovaleraldehyde is commercially obtained by heating D-Glucose with L-Leucine (equation 1).

D-Glucose   L-Leucine   Isovaleraldehyde

Equation 1.

Possibly the best known commercial reaction flavor is for natural 2,5-Dimethyl-3-hydroxy-3 [2H] furanone obtained by heating D-Rhamnose with L-Proline (equation 2).

D-Rhamnose   L-Proline   2,5-Dimethyl-3-hydroxy-
                                        3[2H]furanone

Equation 2.

However, flavorists and chemists alike want to have pure chemicals uncomplicated by the presence of by-products and this has pushed us to the next mechanism of oxazoles, oxazolines, thiazoles, thiazolines and pyrazines from model reactions of carbonyl compounds and ammonium sulfide under low temperature condition. The reactions between α-hydroxyketones (3-hydroxy-2-butanone, 1-hydroxy-2-propanone, 1-hydroxy-2-butanone) and α-dicarbonyls (2,3-butanedione, 2,3-pentanedione) and ammonium sulfide were studied.

Heterocyclic Compound Formation Hypothesis

Takken (2) identified thiazoles and 3-thiazolines from the reaction of 2,3-butanedione and 2,3-pentanedione with ammonia, acetaldehyde and hydrogen sulfide at 20 °C. Study of tetramethylpyrazine (3) also showed that it can be readily formed in 3-hydroxy-2-butanone and ammonia model reaction at 22 °C. Recent study of the model reaction of 3-hydroxy-2-butanone and ammonium acetate at low temperature revealed an interesting intermediate compound, 2-(1-hydroxyethyl)-2,3,4-trimethyl-3-oxazoline, along with 2,4,5-trimethyloxazole, 2,4,5-trimethyl-3-oxazoline, and tetramethylpyrazine were isolated and identified (4,5). We hypothesized that with the introducing of H_2S, replacement of oxygen by sulfur could happen and sulfur-containing heterocyclic compounds such as thiazoles and thiazolines could be formed along with oxazoles, oxazolines and pyrazines.

Experimental Design

To test the volatile heterocyclic compounds formation under low temperature condition, 0.01 mol of one of the following carbonyl compounds including α-hydroxyketones (3-hydroxy-2-butanone, 0.88 g; 1-hydroxy-2-propanone, 0.74 g; 1-hydroxy-2-butanone 0.88g) and α-dicarbonyls (2,3-butanedione, 0.86 g; 2,3-pentanedione 1.00 g) and 0.02 mol of ammonium sulfide (20% wt/wt solution in water) were dissolved in 25 mL distilled water and reacted at 25°C for 2 hours. Immediately follow the reactions, the mixtures were cooled and extracted with methylene chloride three times, dried over anhydrous sodium sulfate, and then concentrated under a stream of N_2 for further GC and GC/MS analyses.

Results and Discussion

Reaction of α-Hydroxyketones (3-Hydroxy-2-butanone, 1-Hydroxy-2-propanone and 1-Hydroxy-3-butanone) with Ammonium Sulfide

In addition to the 3-hydroxy-2-butanone and ammonium acetate model reaction as previously studied *(4,5)*, reaction of 3-hydroxy-2-butanone and ammonium sulfide at 25 °C generated three intermediate compounds including 2-(1-mercaptoethyl)-2,4,5-trimethyl-3-oxazoline, 2-(1-hydroxyethyl)-2,4,5-trimethyl-3-thiazoline and 2-(1-mercaptoethyl)-2,4,5-trimethyl-3-thiazoline in addition to 2-(1-hydroxyethyl)-2,4,5-trimethyl-3-oxazoline. Reaction led to only trace amount of tetramethylpyrazine formation. Previous study of this reaction at six different temperatures (25, 50, 75, 100, 125 and 150 °C) indicated that these intermediate compounds were readily formed at 25 °C and tetramethylpyrazine was the major product at and above 100 °C. Reaction mechanism was discussed in the same paper *(6)*.

In the studies of the reactions of 1-hydroxy-2-propanone and 1-hydroxy-2-butanone and ammonium sulfide, despite four possible intermediates were predicted, only two intermediate compounds were tentatively identified by GC/MS in our study. The mass spectral data of these compounds showed very distinctive base peak pattern (Table I). These results agreed with results from previous study of five different 2-alkyl-2,4,5-trimethyl-2,5-dihydrooxazolines *(7)* which showed same mass spectral fragmentation pattern (m/z = 112) of same types of intermediate compounds. Further study of GC/MS-CI results suggested their molecular weights and as the results, they were tentatively identified as 2-(1-hydroxymethyl)-2,4-dimethyl-3-oxazoline, 2-(1-hydroxymethyl)-2,4-dimethyl-3-thiazoline from 1-hydroxy-2-propanone and ammonium sulfide reaction and 2-(1-hydroxymethyl)-2,4-diethyl-3-oxazoline, 2-(1-hydroxymethyl)-2,4-diethyl-3-thiazoline from 1-hydroxy-2-butanone and ammonium sulfide reaction. GC-MS/EI and GC-MS/CI mass spectra of these compounds and their structures are shown in Figures 1-4.

Proposed reaction mechanism of 1-hydroxy-2-propanone or 1-hydroxy-2-butanone with ammonium sulfide is shown in Figure 5. 2-Iminoalcohols, which were derived from the reaction of ammonia and α-hydroxyketones, were condensed with another molecule of α-hydroxyketone to form 2-(1-hydroxymethyl)-2,4-dialkyl-3-oxazolines. The substitution of -OH group with -SH group resulted in the formation of 2-(1-hydroxyemthyl)-2,4-dialkyl-3-thiazolines. No 2-(1-mercaptomethyl)-substituted isomers were found in either reaction. Upon heating, these intermediates were further converted to the corresponding oxazolines, oxazoles, thiazolines and thiazoles.

Table I. Mass Spectral Base Peaks for Intermediate 3-Oxazolines and 3-Thiazolines Identified in α-Hydroxyketones and Ammonium Sulfide Reactions

Starting α-Hydroxyketone	3-Oxazoline	3-Thiazoline
3-hydroxy-2-butanone	m/z 112	m/z 128
1-hydroxy-2-propanone	m/z 98	m/z 114
1-hydroxy-3-butanone	m/z 126	m/z 142

Reaction of α-Dicarbonyls (2,3-Butanedione and 2,3-Pentanedione) with Ammonium Sulfide

In the reaction of 2,3-butanedione and ammonium sulfide at 25 °C, quantitation of volatile compounds identified in the reaction is summarized in Table II. A pair of interesting intermediate compounds, 5-hydroxy-3-thiazolines, were tentatively identified by GC/MS-EI. 5-Hydroxy-2,4,5-trimethyl-3-thiazolines have been separated and identified by GC/MS-EI and NMR in the early study of volatile flavor compounds of yeast extracts (8). They were also found in the reaction of 2,3-butanedione or 2,3-pentanedione with acetaldehyde, hydrogen sulfide and ammonia (2).

Proposed reaction scheme for 2,3-butanedione and ammonium sulfide is summarized in Figure 6. Since there was no acetaldehyde involved in this model reaction, we proposed that the initial step of cleavage of 2,3-butanedione would required to form an acetaldehyde and an acetamide which was similar to the mechanism described by Grimmett and Richards (9) for the interaction of pyruvaldehyde and ammonia in an ammoniacal environment. Thioacetamide was formed as the result of addition of H_2S to acetamide. In fact, the presence of 1,2,4-trithiolanes and 1,3,5-trithiane in the reaction mixture also explained the possible existence of acetaldehyde. 2,3-Butanedione reacted with H_2S to form an 3-hydroxy-3-mercapto-2-butanone intermediate. It further reacted with an ethylideneamine which derived from the result of acetaldehyde and ammonia interaction. The reaction formed 5-hydroxy-2,4,5-trimethyl-3-thiazoline and further dehydration upon heating led to the formation of 2,4,5-trimethylthiazole.

102

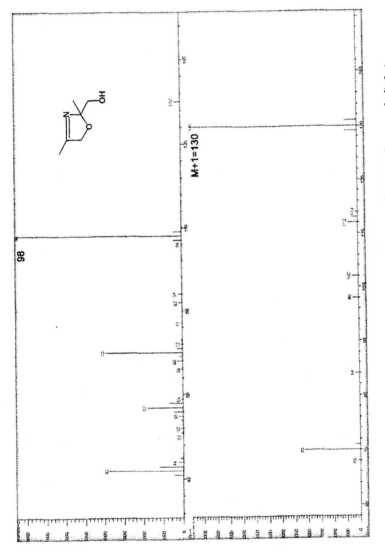

Figure 1: GC-MS/EI and GC-MS/CI mass spectra of 2-(1-hydroxymethyl)-2,4-dimethyl-3-oxazoline

103

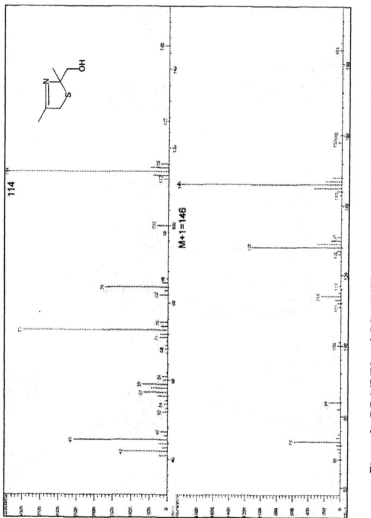

Figure 2: GC-MS/EI and GC-MS/CI mass spectra of 2-(1-hydroxymethyl)-2,4-dimethyl-3-thiazoline

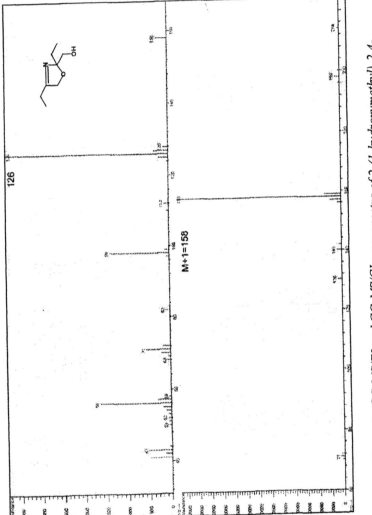

Figure 3: GC-MS/EI and GC-MS/CI mass spectra of 2-(1-hydroxymethyl)-2,4-diethyl-3-oxazoline

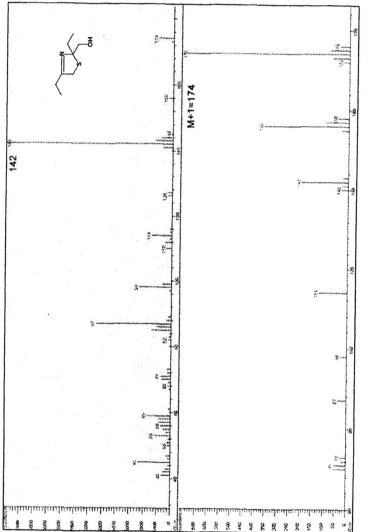

Figure 4: GC-MS/EI and GC-MS/CI mass spectra of 2-((1-hydroxymethyl)-2, 4-diethyl-3-thiazoline.

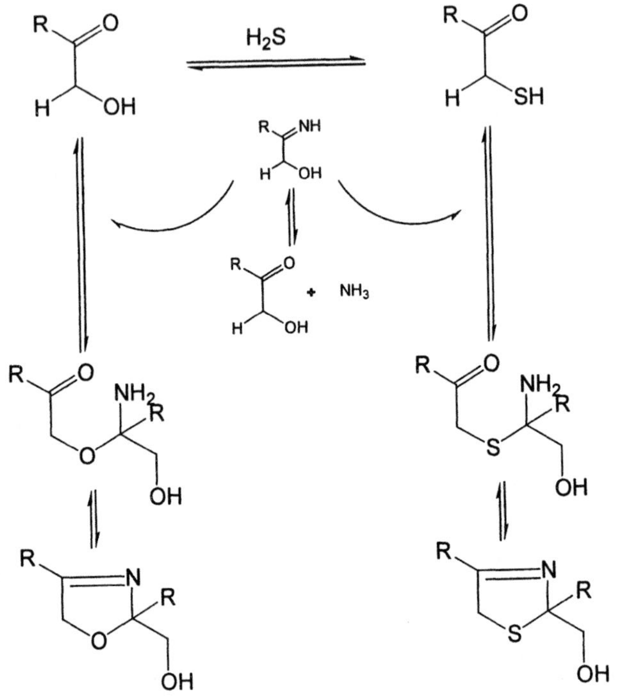

R = CH₃ (1-hydroxy-2-propanone)

R= CH₃CH₂ (1-hydroxy-2-butanone)

Figure 5. Proposed reaction mechanism of 1-hydroxy-2-propanone or 1-hydroxy-2-butanone with ammonium sulfide.

Table II. Compounds identified in the reaction of 2,3-butanedione and ammonium sulfide at 25°C (2 hrs)

	Compounds	Relative amounts (%)*
1	2,3-butanedione	0.01
2	3-hdyroxy-2-butanone	0.01
3	2,4,5-trimethyloxazole	2.79
4	3-mercapto-2-butanone	0.11
5	acetamide	0.52
6	2,4,5-trimethyl-3-thiazoline	trace
7	2,4,5-trimethylthiazole	0.10
8	2,4,5-trimethyl-3-thiazoline	trace
9	tetramethylpyrazine	27.30
10	1,2,4-trithiolane	0.01
11	1,3,5-trithiane	0.01
12	3,5-dimethyl-1,2,4-trithiolane	0.04
13	5-hydroxy-2,4,5-trimethyl-3-thiazoline	0.02
14	3,5-dimethyl-1,2,4-trithiolane	0.04
15	5-hydroxy-2,4,5-trimethyl-3-thiazoline	0.02
16	thioacetamide	0.74
17	2,4,5-trimethylimidazole	4.15

*: As percentage of total chromatogram peak area.

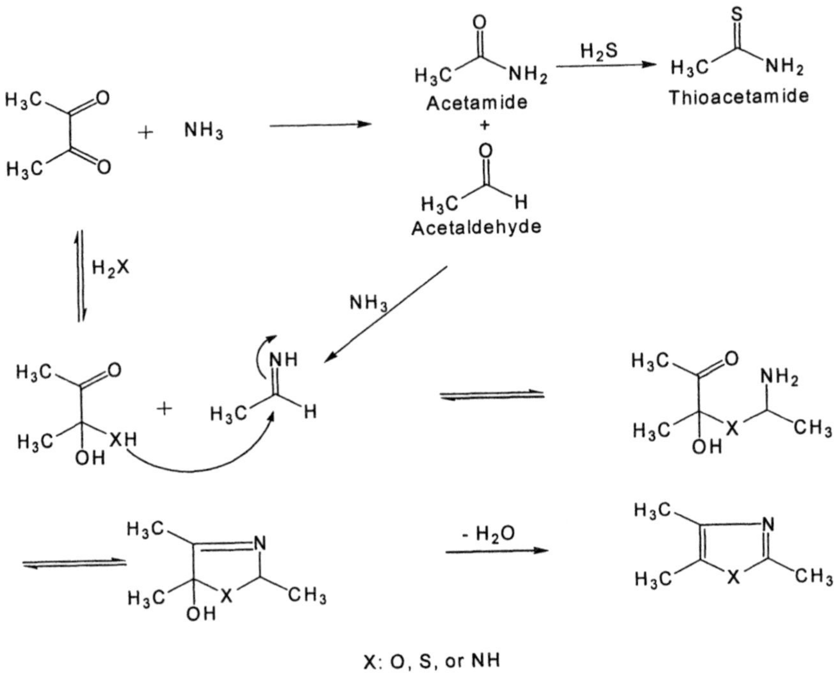

108

Figure 6: Proposed reaction mechanism of 2,3-butanedione with ammonium sulfide at 25 °C.

An interesting end product, tetramethylpyrazine was also presented in the reaction mixture (27.3% of peak area of the total volatiles). Previous study of the model reaction of 2,3-butanedione and ammonium acetate did not yield any tetramethylpyrazine. It is probably due to the reducing environment provided by H_2S which reduced 2,3-butanedione to 3-hydroxy-2-butanone. This explained that both 3-hydroxy-2-butanone and 3-mercapto-2-butanone were found in the reaction mixture. This study also supported the mechanism proposed by Elmore and Mottram (10) who observed that, during the reactions of hydroxyketones with aldehydes and ammonium sulfide, the formation of thiazoles was discouraged due to reducing environment provided by H_2S derived from ammonium sulfide. It is also interesting to note that comparing to previous α-hydroxyketone series tetramethylpyrazine was present at trace levels whereas in the α-dicarbonyl series it was the major product under comparable temperature conditions. The reason for this observed phenomenon is not obvious. It is possible that in the reaction system of acetoin and ammonium sulfide, the concentrations of ammonia and hydrogen sulfide are equal. The hydrogen sulfide is more reactive than ammonia which results in the formation of more sulfur-containing compounds than tetramethylpyrazine compounds. In the system of diacetyl and ammonium sulfide, hydrogen sulfide is used to reduce the diacetyl to acetoin. There is more ammonia than hydrogen sulfide reacting with acetoin, which leads to the formation of tetramethylpyrazine as a major product.

Another major compound, 2,4,5-trimethylimidazole, was also presented in the gas chromatogram (4.15% of peak area of the total volatiles). Shibamoto (11) showed that 2,4,5-trimethylimidazole was the major product of the reaction of diacetyl and ammonium hydroxide at 100 °C for 180 minutes. Formation mechanism might be similar to those of oxazoles and thiazoles with NH$_3$ replacing H$_2$O or H$_2$S (Figure 6).

In the study of 2,3-pentanedione and ammonium sulfide reaction, due to the asymmetric structure of 2,3-pentanedione the formation of four different oxazoles and thiazoles can be expected and indeed were found in the reaction mixture (Figure 7).

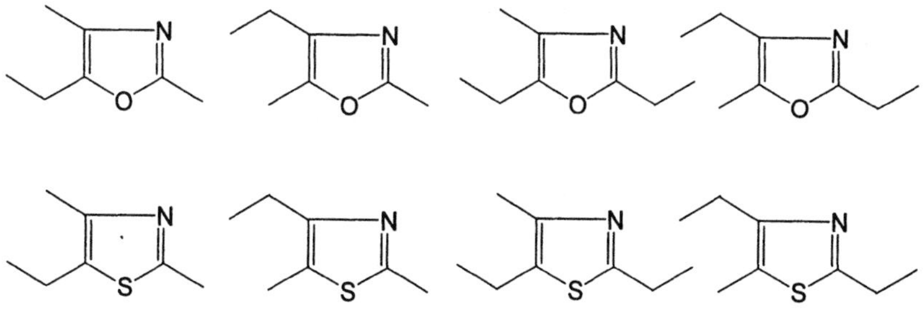

Figure 7. Oxazoles and thiazoles identified in the reaction of 2,3-pentanedione and ammonium sulfide at 25°C.

Conclusion

Model reactions of α-hydroxyketones or α-dicarbonyls and ammonium sulfide at low temperature provided interesting insights of formation of heterocyclic compounds. α-Hydroxyketones reacted with ammonium sulfide through 2-(1-hydroxyalkyl)-3-oxazoline (or thiazoline) intermediates. In the case of α-dicarbonyl compounds, 5-hydroxy-substituted 3-thiazolines were the intermediates and at the same time, under the reducing condition provided by the presence of hydrogen sulfide, α-dicarbonyls were also reduced to hydroxyketones to provide an additional formation routes for oxazoles, oxazolines, thiazoles, thiazolines as well as pyrazines.

References

1. Vernin, G. *The Chemistry of Heterocyclic Flavouring and Aroma Compounds.* Ellis Horwood, Chichester, UK, 1982.
2. Takken, H.J.; van der Linde, L.M.; de Valois, P.J.; van Dort, H.M.; Boelens, M. *In Phenolic, Sulfur, and Nitrogen Compounds in Food Flavors.* Charalambous, G.; Katz, I., Eds.; ACS Symp. Ser. 26, American Chemical Society: Washington, DC, 1976, 114-121.
3. Rizzi, G.P. *J. Agric. Food Chem.* **1988**, *36*, 349-352.
4. Shu, C-K.; Lawrence, B. *J. Agri. Food Chem.* **1995**, *43*, 2922-2924.
5. Fu, H.Y.; Ho, C.T. *J. Agric. Food Chem.* **1997**, *45*, 1878-1882.
6. Xi, J.; Huang, T.C.; Ho, C.-T. *J. Agric. Food Chem.* **1999**, *47*, 245-248.
7. Le Quéré, J.L.; Fournier, N.; Langlois, D.; Henry, R. In: *Heteroatomic Aroma Compounds;* Reineccius, G.; Reineccius, T., Eds., American Chemical Society, Washington DC, 2002, 166-178.
8. Werkhoff, P.; Bretschneider, W.; Emberger, R.; Güntert, M.; Hopp, R.; Köpsel, M. *Chem. Mikrobiol. Technol. Lebensm.* **1991**, *13*, 30-57.
9. Grimmett, M.R.; Richards, E.L. *J. Chem. Soc.* **1965**, 3751-3754.
10. Elmore, J.S.; Mottram, D.S. *J. Agri. Food Chem.* **1997**, *45*, 3595-3602.
11. Shibamoto, T. *J. Appl. Toxic.* **1984**, 97-100.

Chapter 8

Formation of Flavor Compounds by the Reactions of Carbonyls and Ammonium Sulfide under Low Temperature

Junwu (Eric) Xi[1,2] and Chi-Tang Ho[1]

[1]Department of Food Science, Rutgers, The State University of New Jersey, New Brunswick, NJ 08901–8520
[2]Ottens Flavors, 7800 Holstein Avenue, Philadelphia, PA 19153

Aqueous mixtures containing α-hydroxyketones (3-hydroxy-2-butanone, 1-hydroxy-2-propanone, 1-hydroxy-2-butanone) or α-dicarbonyls (2,3-butanedione, 2,3-pentanedione) and ammonium sulfide were reacted at 25 °C for 2 hr. Among the heterocyclic flavor compounds formed were oxazoles, oxazolines, thiazoles, thiazolines and pyrazines. 2-(1-Hydroxyalkyl)-3-oxazolines and 2-(1-hydroxyalkyl)-3-thiazolines were major intermediate compounds identified in α-hydroxyketone systems and on the other hand, 5-hydroxy-3-oxazolines and 5-hydroxy-3-thiazolines were proposed as intermediate compounds in α-dicarbonyl systems.

Carbonyl compounds, ammonia and hydrogen sulfide are some very reactive flavor precursors which could be derived from early stage of Maillard reaction and pre-existing in many food systems. Reactions among them could lead to the formation of various heterocyclic flavor compounds (*1*). However, research work done regarding these reactions were mostly under high temperature conditions. Reaction mechanism under low temperature condition has not been well researched. The purpose of this study was to elucidate formation

111

stage of evolution. The next step in the evolution of reaction flavors goes from amino acids and sugar to the reaction of discrete natural materials with amino acids.

We have reacted natural Diacetyl with natural L-Cysteine to obtain both Tetramethylpyrazine and 3-Mercapto-2-butanone in commercially attractive yields (equation 3).

| Diacetyl | L-Cysteine | | Tetramethyl-pyrazine | 3-Mercapto-2-butanone |

Equation 3.

The pathway for the transformation is believed to involve the formation of an intermediate thione and dihydropyrazine. These materials undergo a mutually beneficial redox reaction to give the observed products. Further evidence is that the yields of products are equal (equation 4).

Equation 4.

Continuing our evolution of reaction flavors brings us to the reaction of materials, which are themselves, isolated from amino acids and sugars to produce unique materials which are frequently very different than the reaction of the parent amino acids and sugar. The reaction of D-Xylose and L-Cysteine is a good example of this principle. A complex meaty tasting mix is obtained from the sugar and amino acid. The mixture contains small amounts of 2-Methyl-3-furanthiol, a potent meat flavor (equation 5).

D-Xylose + L-Cysteine → "Meaty" Mixture Containing 2-Methyl-3-furanthiol

Equation 5.

However, if the L-Cysteine is heated in water, one can obtain both Ammonia and Hydrogen Sulfide (equation 6). D-Xylose can be heated in water to produce Furfural (equation 7).

L-Cysteine $\xrightarrow{\text{Water, Heat}}$ H_2S (Hydrogen Sulfide) + NH_3 (Ammonia)

Equation 6.

D-Xylose $\xrightarrow{\text{Water, Heat}}$ Furfural

Equation 7.

If the Hydrogen Sulfide and Furfural are allowed to react in a buffered ethanolic solution at -35°C, substantial yields of Difurfuryl Disulfide (FEMA 3146) are obtained. This is a powerful flavor material with a roasted coffee and roasted meaty taste. It is widely used to obtain roasted notes in meat flavors (equation 8).

Difurfuryl Disulfide
FEMA 3146

Flavor Descriptors: Roasted Coffee, Meaty
Occurrence: Coffee, Bread
Flavor Threshold: Estimated @ 0.05 ppb

Equation 8.

We believe the mechanisms of the reaction are as shown in equation 9. The final step is an unusual one. The thiofurfural intermediate undergoes a redox reaction with Hydrogen Sulfide to produce Difurfuryl Disulfide and Sulfur (equation 9). The reaction is carried out in a Nitrogen atmosphere, which would preclude the facile oxidation of Furfuryl Mercaptan to Furfuryl Disulfide.

Equation 9.

Some of these reaction flavors occur in nature. The oxidation of mercaptans to form disulfides and other products is in part responsible for the loss of flavor from fresh roasted coffee only a short time after brewing. Furfuryl Mercaptan is one of the most potent flavor components in coffee and is responsible for a fresh brewed coffee flavor. While it is present at only 1.55 ppm versus Furfuryl Alcohol (515 ppm) and Furfural (157 ppm), its flavor contribution is 310,014 times its flavor threshold (0.005 ppb). If there were a 100% conversion of Furfuryl Mercaptan to Difurfuryl Disulfide, the coffee would appear to be 10 times weaker due to Difurfuryl Disulfide's flavor threshold being 0.05 ppb, or ten times weaker than Furfuryl Mercaptan (equation 10). The overall effect of this oxidation is that the coffee has lost its flavor.

Flavor Descriptors: Powerful Burnt, Coffee, Meaty
Occurrence: Coffee, Cooked Beef
Flavor Threshold: 0.005 ppb

Furfuryl Mercaptan
FEMA 2493

Difurfuryl Disulfide
FEMA 3146

Flavor Descriptors: Roasted Coffee, Meaty
Occurrence: Coffee, Bread
Flavor Threshold: Estimated @ 0.05 ppb

Equation 10.

The oxidation of mercaptans can be useful to prepare other flavor molecules. Furfuryl Mercaptan and Methyl Mercaptan can be oxidized together to give Methyl Furfuryl Disulfide, a potent material useful for bread, pork and other meat products. It is also the third most active flavor material in fresh brewed coffee, being present at 0.38 ppm or 9,623 times its flavor threshold (0.04 ppb) (equation 11).

Methyl Furfuryl Disulfide
FEMA 3362

Flavor Descriptors: Roasted, Bread Crust
Occurrence: Coffee, Cooked Pork, Wheat Bread
Flavor Threshold: 0.04 ppb

Equation 11.

Recent reaction flavor work in our labs has lead to a recently GRAS listed material, natural 4-Mercapto-4-methyl-2-pentenone (FEMA 3997), from the reaction of natural Hydrogen Sulfide and natural 4-Methyl-3-penten-2-one (equation 12). While the flavor descriptors in the literature vary from catty, buchu, black currant, broom tree to cassis, we have found the material to have a meaty, Chinese pork type of flavor when used in savory applications.

4-Methyl-3-penten-2-one

4-Mercapto-4-methyl-2-pentanone
FEMA 3997

Flavor Descriptors: Catty, Buchu, Black Currant,
Broom Tree, Chinese Pork, Cassis
Occurrence: Grapefruit, Cassis, Wine
Flavor Threshold: 2×10^{-5} ppb

Equation 12.

Similar reactions have been used to generate pleasant pork notes from the reaction of natural Isovaleraldehyde and natural Ammonium Sulfide, as shown in the following figure. The chemistry of this pork flavor formulation has been the subject of several patents[2][3] and numerous publications (equation 13).

Equation 13.

The sheer volume of material written certainly testifies to the commercial and intellectual importance of these "next generation" reaction flavors.

References

[1] Buckholz, L., Byrne, B. and Sudol, M. U.S. Patent 4,882,184, 1989.

[2] Mookerjee, B. D., Shu, C. and Vock, M. H. U.S. Patent 4,200,742, 1980.

[3] Wiener, C. U.S. Patent 3,650,771, 1972.

Chapter 9

Amadori Compounds of Cysteine and Their Role in the Development of Meat Flavor

Kris B. de Roos, Kees Wolswinkel, and Gerben Sipma

Givaudan Nederland BV, P.O. Box 414, 3770 AK Barneveld,
The Netherlands

In search for the key intermediates to meat flavor development in heated cysteine-sugar systems, it was found that in addition to thiazolidine derivatives also the tetrahydro-1,4-thiazine derivatives are being formed. These tetrahydro-1,4-thiazines, which are the cyclic form of Amadori compounds of cysteine, have excellent meat flavor precursor properties and are likely to play a prominent part in meat flavor development. Another major pathway to meat flavor development is the reaction of cysteine with the Amadori compounds of other amino acids. Model experiments showed that both pathways are probably of about equal importance for flavor development in boiled meat and process flavorings, this in spite of the low reactivity of cysteine with sugars. It seems that the first pathway is general-acid-catalyzed by the other amino acids, whereas the second pathway is inhibited by cysteine.

Introduction

Cysteine is an important precursor of meat flavor and is therefore often being used in precursor systems for the industrial production of meat process flavorings (1-4). Meat flavor development in these systems is usually based on the Maillard reaction of cysteine (and other amino acids) with sugars. Unfortunately, there are a few complications that prevent that high yields of volatile flavor compounds are obtained from these reactions. The first complication is the low stability of the generated flavor compounds, which leads to the development of off-flavors during heating and storage (5, 6). The second complication is the inhibition of the Maillard reaction by cysteine. This inhibitory effect of cysteine is attributed to the formation of thiazolidines, (hemi)thioacetals and -ketals upon reaction with sugars and other (di)carbonyl compounds (5, 7). Due to the low reactivity of cysteine in its reaction with sugars (8), it is not easily possible to generate high concentrations of labile flavor compounds under the mild conditions that are required to prevent their decomposition.

Previous work has demonstrated that sugars that show highest conversion to thiazolidines in their reaction with cysteine produce also most browning and flavor (5). This looks contradictory since the formation of thiazolidines is seen as the cause of browning inhibition. To explain these results, it was assumed that the pathways to thiazolidine formation as well as to browning and flavor formation proceed via the same intermediate, which is the Schiff's base of cysteine with the carbohydrate.

The objective of the present work was to learn more about the initial stages of the reaction between cysteine and sugars to meat flavor compounds. This information can provide a clue to a more efficient conversion of the precursors into meat flavor compounds.

Experimental procedures

Materials

L-cysteine, L-cysteine.HCL monohydrate and D-xylose were obtained from Diamalt AG, Munich, Germany. Other chemicals were from Merck, Darmstadt, Germany. Ion exchange resins were from BioRad Laboratories B.V., Veenendaal, Netherlands.

HPLC analysis

Analyses were carried out on a 300 x 4.6 mm i.d. column of Lichrosorb NH_2 (10 μm particle size). Detector: refractive index. Mobile phase: 0.02 M ammonium formate in acetonitrile-water 60:40 adjusted to pH 6 with formic acid. Flow rate: 2 ml/min. For quantification, pure compounds were used as external standard.

Spectroscopy

Infrared spectra were obtained in KBr using a Perkin Elmer 225 Grating Spectrometer. NMR spectra were recorded on a Jeol FX-100, operating at 99.55 MHz for proton NMR and at 25.0 MHz for ^{13}C NMR. The compounds were dissolved in D_2O. Internal reference: sodium 3-trimethylsilyl [2,2,3,3-$^2H_4$]-propionate.

Reactions of cysteine with sugars – general procedure

L-cysteine.HCl monohydrate (17.6 g; 0.1 mole) was dissolved in 50 ml of demineralized water contained in a 150-ml round-bottomed flask. Optionally, disodium hydrogen phosphate (14.2 g; 0.10 mole) or 90% lactic acid (10 g; 0.10 mole) was added and the pH of the resulting solution adjusted to 5.0 by slowly adding with stirring an aqueous 50% sodium hydroxide solution. The sugar (0.11 mole) was then added and the mixture stirred for 1 h at 50 °C to obtain an equilibrium mixture of thiazolidine, cysteine and sugar. After cooling to room temperature the pH was adjusted to the desired value by addition of 50% sodium hydroxide solution or concentrated hydrochloric acid. If necessary, the water content was adjusted by evaporation *in vacuo* or the addition of water. The flask was then equipped, via a double neck adapter, with a mechanical stirrer and a reflux condenser, and heated with stirring in a thermostated water bath for 2-8 h at 85-95 °C. The reaction was monitored by HPLC analysis of samples taken at regular time intervals.

Isolation of a ketose-L-cysteine

After completion of the cysteine-sugar reaction the mixture was diluted with water to a solids content of about 20% and sent through a column of Bio-Rad AG 50W-X4 (H^+) cation exchange resin of 200-400 mesh. The column was washed with two column volumes of demineralized water and the adsorbed substances were displaced from the column with 0.3 M ammonia. The thiazolidine came off the column first followed by the Amadori compound and cysteine. The fraction containing the Amadori compound was concentrated in vacuo to dryness. The residue was dissolved in a small volume of water and just enough ethanol was added to cause cloudiness. The Amadori compound separated slowly in the form of a colorless crystalline product.

D-xylulose-L-cysteine: Prepared from D-xylose. M.p.: decomp. > 170°C.
^{13}C NMR: δ/ppm, major isomer/minor isomer: 26.9 (t)/25.6 (t) (C_6); 53.8 (t)/50.2 (t) (C_3); 61.5 (d)/57.0 (d) (C_5); 65.2 (t)/65.2 (t) (CH_2OH); 71.9 (d)/71.6

(d) (CHOH); 75.9 (d)/76.6 (d) (CHOH); 81.5 (s)/80.9 (s) (C_2); 173.4 (s)/173.0 (s) (C_7).
Found   : C 37.87; H 5.96; N 5.51; S 12.65; O 37.80.
Calc. for $C_8H_{15}NSO_6$ (253.27): C 37.94; H 5.97; N 5.53; S 12.66; O 37.90.
The product was dried prior to element analysis (*in vacuo* over P_2O_5, 18 h at 69°C). No crystal water was present (Karl-Fisher).

D-*ribulose-L-cysteine*: Prepared from D-ribose. M.p.: 132-135°C (decomp.).
^{13}C NMR: δ/ppm, major isomer/minor isomer: 26.8 (t)/25.7 (t) (C_6); 55.2 (t)/51.1 (t) (C_3); 61.9 (d)/57.3 (d) (C_5); 65.6 (t)/65.3 (t) (CH_2OH); 74.0 (d)/73.4 (d) (CHOH); 75.9 (d)/77.8 (d) (CHOH); 82.7 (s)/81.1 (s) (C_2); 174.0 (s)/173.5 (s) (C_7).

L-ribulose-L-cysteine: Prepared from L-arabinose. M.p.: 171-173°C (decomp.).
^{13}C NMR: δ/ppm, major isomer/minor isomer: 26.9 (t)/25.8 (t) (C_6); 54.9 (t)/51.7 (t) (C_3); 62.1 (d)/57.2 (d) (C_5); 65.4 (t)/65.6 (t) (CH_2OH); 73.7 (d)/73.9 (d) (CHOH); 77.7 (d)/76.2 (d) (CHOH); 81.1 (s)/82.8 (s) (C_2); 174.0 (s)/173.6 (s) (C_7).

D-*fructose-L-cysteine*: Prepared from D-glucose. M.p.: 175-177°C (decomp.).
^{13}C NMR: δ/ppm, major isomer/minor isomer: 27.1 (t)/25.9 (t) (C_6); 54.1 (t)/50.3 (t) (C_3); 61.6 (d)/57.2 (d) (C_5); 65.1 (t)/65.1 (d) (CH_2OH); 71.6 (d)/71.3 (t) (CHOH); 73.3 (d)/73.5 (d) (CHOH); 74.4 (d)/76.1 (d) (CHOH); 81.8 (s)/81.1 (s) (C_2); 173.6 (s)/173.0 (s) (C_7).

6-*deoxy-L-fructose-L-cysteine*: Prepared from L-rhamnose. M.p.: decomp. > 183°C. ^{13}C NMR: δ/ppm, major isomer/minor isomer: 21.1 (q)/21.2 (t) (CH_3); 26.9 (t)/25.9 (t) (C_6); 54.3 (t)/50.8 (t) (C_3); 62.1 (d)/57.0 (d) (C_5); 69.9 (d)/69.7 (d) (CHOH); 75.2 (d)/74.7 (d) (CHOH); 75.8 (d)/75.4 (d) (CHOH); 81.3 (s)/81.8 (s) (C_2); 173.9 (s)/173.5 (s) (C_7).

Continuous process for the production of D-xylulose-L-cysteine

The continuous reactor consisted of a 45x2.5 ID cm glass column with temperature control jacket packed with Bio-Rex 9 anion exchange resin (OH⁻), 100-200 mesh. A restriction consisting of a 500 x 0.3 mm ID Teflon tubing was connected to the column outlet to prevent the formation of gas bubbles in the column. The column was thermostated at 90 °C. A 50% thiazolidine solution, prepared by heating equimolar amounts of cysteine and xylose in water for 1 h at 50 °C, was pumped through the column at a rate of 0.5 ml/min. Analysis of samples taken at regular time intervals showed that the conversion of thiazolidine to Amadori compound increased from12% after 2 hrs to 35% after 6 hrs. The yield of the corresponding batch process is 22.5%.

Results and Discussion

Liquid chromatographic analysis of the changes occurring during heating of a concentrated aqueous solution of cysteine and xylose showed that the initially high concentrations of the two 2-xylulosylthiazolidine-4-carboxylic acid isomers (peak **d** in Figure 1) decreased slowly, whereas at the same time a new unknown compound was formed (peak **?** in Figure 1). After a reaction time of 2 hours the concentration of this compound achieved a maximum and then decreased again.

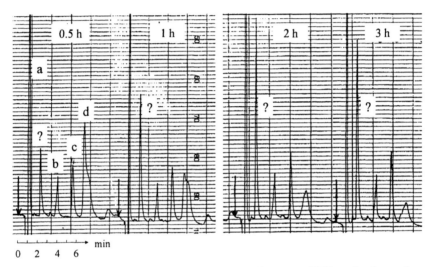

Figure 1. Liquid chromatograms of a mixture of cysteine.HCL and xylose after different periods of heating at 90 °C and pH 5.0 (lactate buffer). Moisture content: 20%. a: solvent + xylose + cysteine; ?: unknown; b: chloride; c: lactate; d: 2-xylulosylthiazolidine-4-carboxylic acid (compound 7 in Figure 2; 2 isomers).

The unknown compound was isolated by means of ion displacement chromatography. A crystalline product was obtained which was identified as 2-hydroxy-2-(1'(S), 2'(R), 3'-trihydroxypropyl)-tetrahydro-2H-1,4-thiazine-5(S)-carboxylic acid (Figure 2, compound **9a**, n=3)). This tetrahydro-1,4-thiazine is the cyclic hemithioacetal form of the Amadori compound of cysteine and xylose (D-xylulose-L-cysteine). In water, the Amadori compounds of cysteine (ketose-L-cysteines) exist in two isomeric forms with the oxygen atom at C_2 in either cis or trans position to the carboxyl group at C_5. On the basis of NMR analysis, element analysis and acetylation experiments, it was concluded that these Amadori compounds occur predominantly in the monocyclic form **9a**.

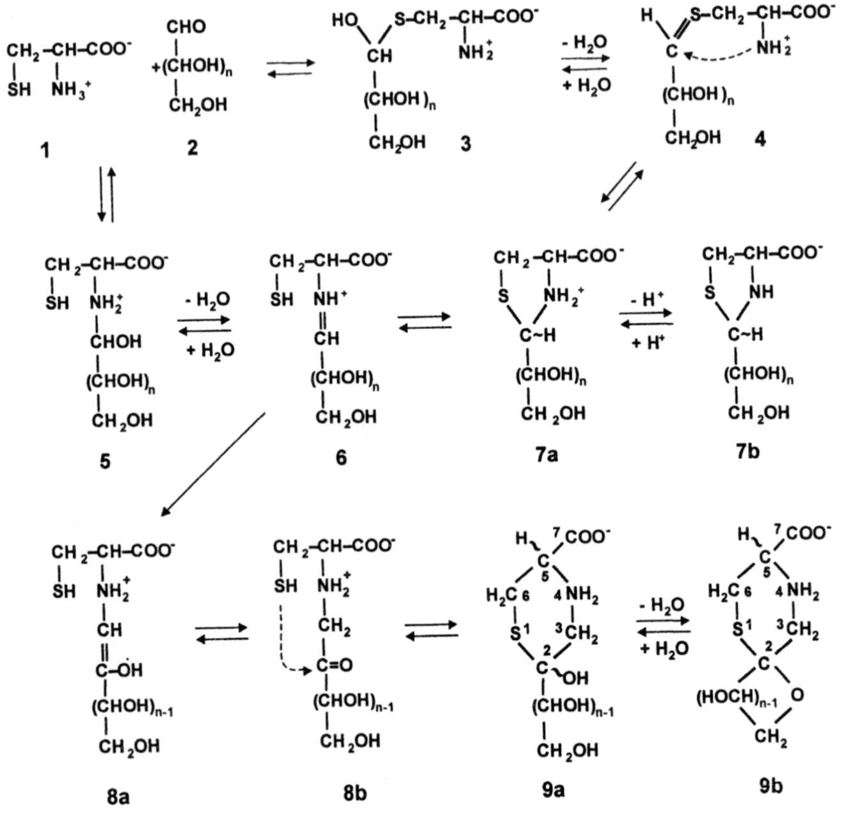

Figure 2 (chemical scheme):

Compound **1**: CH$_2$–CH–COO$^-$ / SH NH$_3^+$

Compound **2**: CHO / +(CHOH)$_n$ / CH$_2$OH

Compound **3**: HO–CH(S–CH$_2$–CH–COO$^-$ / NH$_2^+$) / (CHOH)$_n$ / CH$_2$OH

Compound **4**: H–C(S–CH$_2$–CH–COO$^-$ / NH$_2^+$) / (CHOH)$_n$ / CH$_2$OH   ($-H_2O$ / $+H_2O$)

Compound **5**: CH$_2$–CH–COO$^-$ / SH NH$_2^+$ / CHOH / (CHOH)$_n$ / CH$_2$OH   ($-H_2O$ / $+H_2O$)

Compound **6**: CH$_2$–CH–COO$^-$ / SH NH$^+$ / CH / (CHOH)$_n$ / CH$_2$OH

Compound **7a**: CH$_2$–CH–COO$^-$ / S NH$_2^+$ / C~H / (CHOH)$_n$ / CH$_2$OH   ($-H^+$ / $+H^+$)

Compound **7b**: CH$_2$–CH–COO$^-$ / S NH / C~H / (CHOH)$_n$ / CH$_2$OH

Compound **8a**: CH$_2$–CH–COO$^-$ / SH NH$_2^+$ / CH / C–OH / (CHOH)$_{n-1}$ / CH$_2$OH

Compound **8b**: CH$_2$–CH–COO$^-$ / SH NH$_2^+$ / CH$_2$ / C=O / (CHOH)$_{n-1}$ / CH$_2$OH

Compound **9a**: H,^{7}COO$^-$ C$_5$ / H$_2$C$_6$ ^{4}NH$_2$ / S^1 ^{3}CH$_2$ / ^{2}C–OH / (CHOH)$_{n-1}$ / CH$_2$OH   ($-H_2O$ / $+H_2O$)

Compound **9b**: H,^{7}COO$^-$ C$_5$ / H$_2$C$_6$ ^{4}NH$_2$ / S^1 ^{3}CH$_2$ / ^{2}C / (HOCH)$_{n-1}$ O / CH$_2$

Figure 2. Proposed mechanisms of thiazolidine and tetrahydro-1,4-thiazine formation in cysteine-sugar systems (on the analogy of Kallen (9) and Hodge (10)).

A proposed mechanism for the formation of the Amadori compounds of cysteine is shown in Figure 2. In this scheme the Schiff's base **6** is the key intermediate to the formation of both the thiazolidine and the Amadori compound. At pH < 7, the route to Amadori compounds is preferred. This is concluded from the observation that the most reactive sugars (those that show highest conversion to condensation products **6** and **7**) show also the highest rate of browning and flavor development (Figure 3a). Apparently, the equilibrium between compounds **6** and **7** lies then far enough to the side of Schiff's base **6** to allow its rapid conversion to Amadori compound **8**, which is the key intermediate to brown pigments and flavor compounds. At pH > 9, however, the opposite occurs: the reactions with the most reactive sugars show then highest inhibition due to a shift of the equilibria to the side of the strongest acid, which is the thiazolidinecarboxylic acid **7** (Figure 3b).

The Amadori compounds of cysteine possess excellent meat flavor precursor properties. This holds in particular for those ketose-L-cysteines that are derived from aldopentoses. Ketose-L-cysteines derived from aldohexoses have also meat flavor precursor properties but are less reactive and produce a different meaty flavor profile. The precursor properties of the Amadori compounds are superior to those of the corresponding thiazolidines or equivalent cysteine-sugar mixtures. Most advantage from the use of Amadori compounds is obtained in relatively dilute aqueous media where the formation of Amadori compounds is the rate-determining step. In canned soups and meats, the intensity of the generated meat flavor is about 10 times higher than the intensity generated from the equimolar cysteine-sugar mixtures.

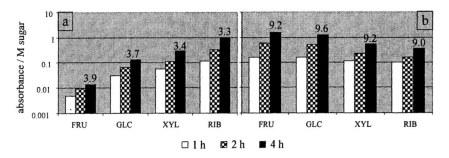

Figure 3. Browning in cysteine-sugar systems in dilute aqueous medium (74-79% water) at 90 °C. a: equimolar mixtures of cysteine and sugar. b: equimolar mixtures of cysteine, sugar and NaOH (5).

Among the major volatile sulfur compounds generated during the preparation of the ketopentose-L-cysteines are 2-methylfuran-3-thiol, 2-furfurylthiol and 3-mercapto-2-pentanone (*13*, Table I). These compounds have also been identified as character impact compounds of boiled meat flavor (*11, 12*). This suggests that ketopentosese-L-cysteines are playing a role in the development of boiled meat flavor. Ketohexose-L-cysteines do not produce these character impact compounds in significant quantities (Table I) and are therefore unlikely to play a role in boiled meat flavor development.

Table 1. Likens-Nickerson extraction of mixtures of cysteine with carbonyl compounds under optimum conditions for Amadori compound formation [a]

	Quantity produced (mg/mole of cysteine)			
	Ribose	Xylose	Glucose	Norfuraneol
2-Methylfuran-3-thiol	43.5	36	< 0.05	0.9
2-Furfurylthiol	2.8	3.3	< 0.05	10.4
3-Mercapto-2-pentanone	8.0	5.6	0.5	< 0.05
2-Mercapto-3-pentanone	16.4	10.5	< 0.05	24.3

[a] Cysteine: carbonyl compound : lactate = 1 : 1 : 1.67; water content: 37% (w/w); pH 5.0; boiling for 1.5 h.

It is surprising that 4-hydroxy-2,5-dimethyl-3(2H)-furanone (norfuraneol), which is often proposed as a key intermediate in the reaction of cysteine with pentoses to meat flavor compounds (14-18), does not produce higher yields of 2-methylfuran-3-thiol and 3-mercapto-2-pentanone during boiling than ribose. This indicates that norfuraneol is not playing a significant role in the formation of these impact compounds during boiling of meat. This conclusion is confirmed in recent work of Cerny and Davidek (19), who used similar heating conditions. It seems that the reaction of norfuraneol with cysteine or H_2S results only in high amounts of 2-methylfuran-3-thiol and 3-mercapto-2-pentanone when heated in aqueous solutions at 140 °C or higher in a closed system (14-18). Such conditions are not representative for conditions normally being used in the kitchen during the preparation of meat for consumption.

The role of Amadori compounds of cysteine in meat flavor development

In mixtures of amino acids and sugars, meat flavor can be developed via different pathways. Cysteine does not only generate meat flavor via keto-L-cysteines but it reacts also with (the decomposition products of) the Amadori compounds of other amino acids to produce meat flavors that are of comparable intensity (5). This means that there are at least two major pathways to meat flavor development in such Maillard reaction systems:

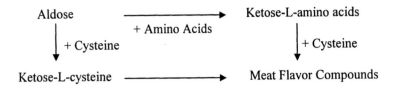

An impression of the relative importance of both routes was obtained by studying the Amadori compound formation in model systems consisting of glucose, cysteine and another amino acid (Table II). It appeared that in a concentrated cysteine-glycine-glucose system (15% water; no pH adjustment, initial pH 5-6), the Amadori compounds of cysteine and glycine are formed at about the same rate. This is remarkable because glycine is about 100 times more reactive than cysteine in the Maillard reaction (8). A possible explanation is that the reaction of cysteine is general-acid-catalyzed by glycine, whereas the reaction of glycine is inhibited by cysteine. In systems where the glycine is replaced by the less reactive L-proline, the formation of D-fructose-L-cysteine proceeds at a clearly higher rate than the formation of D-fructose-L-proline.

Table II. Formation of Amadori compounds in cysteine-amino-acid-glucose model systems at 90 °C.

Composition of model system				Relative HPLC peak areas		
CysSH moles	Glc Moles	Amino acid (moles)	Water w%	Reaction time (h)	D-Fru-L-Cys	D-Fru-AA [a]
1	1.5	Glycine (1)	15	1	26	29
				2	33	29
1	1.5	L-Proline (1)	15	1	40	6
				2.5	46	16
1	1	Glycine (1)	79	2	1.5	0.3
				4	4	1
1	1	L-Proline (1)	79	2	3.8	0.1
				4	7.5	0.25

[a] D-Fru-AA = Amadori compound of glucose with amino acid other than cysteine.

If the medium becomes more dilute, the rate of Amadori compound formation is in general drastically reduced. However, this does not apply to the cysteine reaction. As a consequence, cysteine becomes clearly the most reactive amino acid in dilute neutral media (Table II). A possible explanation for the relatively high reactivity of cysteine in dilute aqueous systems is the high concentration of the cysteine-sugar condensation products thanks to the assistance by the thiol group. Therefore, in dilute systems the concentration of the Schiff's base of cysteine **6** will be much higher than the concentrations of the corresponding Schiff's bases of the other amino acids.

In real meat and often also in precursor systems for meat process flavorings, the concentration of free cysteine is much lower than the total concentration of all other free amino acids together (20-22). This means that in spite of the often much higher reactivity of cysteine in these systems, a major part of the meat flavor will be developed via the Amadori compounds of the other amino acids, in particular, in concentrated media such as crusts. So, where the Amadori compounds of cysteine might still play the major role in boiled meat flavor development, it is possible that in the concentrated medium used for the preparation of meat process flavorings the other pathway to meat flavor development is higher importance.

If cysteine is indeed an important precursor of meat flavor, it is at first glance difficult to understand why similar concentrations of cysteine in beef, pork and chicken (20-22) produce such different concentrations of thiols during boiling (23). A possible explanation for this apparent anomalous result is that the aldehydes, which are easily formed during heating of chicken and other meats with relatively high concentrations of unsaturated fats, bind the cysteine in the form of thiazolidine derivatives thus making the cysteine unavailable for reaction with sugars. Moreover, the (unsaturated) aldehydes can bind the thiols generated from cysteine to form (hemi)thioacetals and sulfides. These reactions might also be responsible for the low concentrations of free cysteine found in meat extracts.

There is still another pathway along which cysteine can produce the meaty sulfur compounds and that is its reaction with thiamine. This reaction, when carried out under the conditions of Table I, generates about 25 times more 2-methylfuran-3-thiol than the reaction of cysteine with ribose. Nevertheless, it was found by Grosch et al. (24) that the addition of the cysteine-thiamine system to meat does not enhance the formation of 2-methylfuran-3-thiol during boiling. It seems that in the highly diluted medium of boiled meat, the rate of this second-order reaction is too slow to produce significant amounts of volatile thiols. The first-order reaction of ketose-cysteines to flavor compounds, on the other hand, is not influenced by dilution and might therefore have a better chance to contribute significantly to boiled meat flavor development. Of course, the ketose-L-cysteines have first to be formed but this seems to be feasible because the formation of these Amadori compounds is not so strongly influenced by dilution (see Table II). However, more work is necessary to find out which pathways are contributing to the formation of the meaty sulfur compounds under the different conditions of meat processing.

Maximizing flavor yields in cysteine-sugar systems

One of the major problems encountered in maximizing the flavor yields from cysteine-sugar systems is the high instability of the thiols that are generated. The thiols do not only react with melanoidins but also with each other (6, 25). The result is that the characteristic meaty flavor disappears and a rubbery off-flavor is formed. Since the rates of the second-order reactions between the thiols are proportional to the square of the thiol concentrations, it is clear that products with high concentrations of thiols will become increasingly unstable. Maximization of flavor yields in cysteine-sugar systems is therefore difficult to reconcile with optimum flavor profile.

A practical approach to the production of more stable meat process flavorings consists of heating cysteine and sugar together with other, preferably nature-identical, meat flavor precursors. The resulting dilution of the cysteine-sugar system reduces the risk of reactions between the thiols. This risk is further reduced by the occurrence of competing biomimetic reactions of the thiols with other compounds. Both factors result in higher flavor stability and a more complete meat flavor profile. A practical approach to the preparation of process flavorings of high flavor strength consists of heating cysteine with (a mixture of) Amadori compounds of other amino acids (5).

Another approach to overcome the flavor instability problem is the *in situ* generation of the unstable flavor compounds just prior to consumption of the product. In principle, this is possible with ketopentosese-L-cysteines as the flavor precursors. Unfortunately, this approach is frustrated by the high costs of the pure Amadori compounds and their relatively low reactivity in products under kitchen conditions (Figure 4).

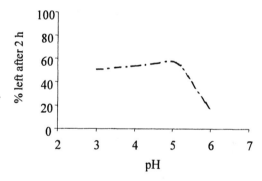

Figure 4. Effect of pH on the decomposition of D-xylulose-L-cysteine in a boiling 0.5 M aqueous phosphate buffer.

The high costs of the ketose-L-cysteines are due to low yields and a time-consuming purification. Under optimum conditions (1:1 molar ratio; water content 30-40%; buffer pH 4-5), the maximum yield of D-xylulose-L-cysteine is only 35%, which is due to its instability under the conditions of its formation. Use of a continuous reactor in which the residence time of the Amadori compound is less than that of the thiazolidine allowed 50% higher conversions to be obtained under similar reaction conditions. Unfortunately, the special requirements for a continuous process prevented the use of the optimum

128

conditions for Amadori compound formation so that the maximum yields were again not higher than 35%.

In theory, a crude cysteine-sugar reaction product with a high content of Amadori compound can also be used as a flavor precursor system. However, to avoid off-flavor formation, the cysteine-xylose reaction has already to be discontinued before 75% of the maximum Amadori compound concentration has been achieved. Moreover, dilution with other precursors and/or a flavor carrier is necessary for acceptable stability. In some cases, the use of mixtures of cysteine with Amadori compounds of other amino acids (5) can also be a suitable alternative for the *in situ* generation of meat flavors.

References

1. Van den Ouweland, G.A.M.; Olsman, H.; Peer, H.G. In: *Agricultural and Food Chemistry; Past, Present, Future*; Teranishi, R., Ed.; Avi Publishing: Westport, Conn., 1978, pp 292-314.
2. May, C.G. *Food Trade Rev.*, **1974**, *44*, 7-14.
3. Wilson, R.A. *J. Agric. Food Chem.* **1975**, *23*, 1032-1037.
4. MacLeod, G. *Developments in Food Flavors*; Birch, G.G.; Lindly, M.G., Eds.; Elsevier: London, **1986**, pp 191-223.
5. De Roos, K.B. *Thermal and Enzymatic Conversions*; Teranishi, R.; Takeoka, G.R.; Güntert, M., Eds.; ACS Symposium Series 490; American Chemical Society: Washington, DC. 1992, pp 203-216.
6. Van Seeventer, P.B.; Weenen, H.; Winkel C.; Kerler, J. *J. Agric. Food Chem.* **2002**, *49*, 4292-4295.
7. Molnar-Perl, I.; Friedman, M. *J. Agric Food Chem.* **1990**, *38*, 1648-1651.
8. Ames, J.M. *Chemistry and Industry*, **1986**. 362-363.
9. Kallen, R.G. *J. Amer. Chem. Soc.* **1971**, *93*, 6236.
10. Hodge, J.E. In: *Chemistry and Physiology of Flavors*; Schultz, H.W.; Day, E.A.; Libbey, L.M., Eds.; Avi Publishing: Westport, Conn., 1967, pp 465-491.
11. Gasser, U.; Grosch, W. *Z. Lebensm. Unters. Forsch.* **1990**, *190*, 3-8.
12. Kerscher, R.; Grosch, W. *Z. Lebensm. Unters. Forsch. A* **1997**, *204*, 3-6.
13. Y. Laarhoven, Givaudan. Unpublished results.
14. Van den Ouweland, G.A.M.; Peer H.G. *J. Agric Food Chem.* **1975**, *23*, 501-505.
15. Shu, C-K.; Ho, C-T. *J. Agric. Food Chem.* **1988**, 36, 801-803.
16. Hofmann, T.; Schieberle P. *J. Agric. Food Chem.* **1998**, *46*, 235-241.
17. Whitfield, F.; Mottram, D. *J. Agric. Food Chem.* **1999**, *47*, 1626-1634.
18. Whitfield, F.; Mottram, D. *J. Agric. Food Chem.* **2001**, *49*, 816-822.
19. Cerny, C.; Davidek, T. *J. Agric. Food Chem.* **2003**, *51*, 2714-2721.

20. Macy, R.L.; Naumann, D.H.; Bailey, M.E. *J. Food Sci.* **1964**, *29*, 142-148.

21. Nishimura, T.; Rhue, M.R.; Okitani, A.; Kato, H. *Agric. Biol. Chem.* **1988**, *52*, 2323-2330.

22. Aliani, M.; Farmer, L.J. Hagan, T.D.J. In: *Flavor Research at the Dawn of the Twentieth-first Century*; Le Quéré J.L.; Etiévant P.X., Eds.; Lavoisier, Cachan: France, 2003, pp 220-223.

23. Kerscher, R.; Grosch, W. In: *Frontiers of Flavour Science*; Schieberle, P.; Engel, K.-H., Eds.; Deutsche Forschungsanstalt für Lebensmittelchemie: Garching, Germany, 2000, pp 17-20.

24. Grosch, W.; Zeiler-Hilgart, G.; Cerny, C.; Guth, H. In: *Prog. Flavor Precursor Stud. Proc. Int. Conf.*; Schreier, P.; Winterhalter, P., Eds.; Allured Publishing: Carol Stream II, 1993, pp 329-342.

25. Hofmann, T.; Schieberle, P. *J. Agric. Food Chem.* **2002**, *50*, 319-326.

Chapter 10

Cystine and Cysteine: Now Available via Vegetarian Fermentation

Christoph Winterhalter

Wacker Chemie GmbH, Hanns Seidel Platz 4, 81737 Munich, Germany
(www.wacker.com)

Abstract

Due to a new patented fermentation process it is now possible
to get access to natural vegetarian cystine and cysteine. The
classical route to these amino acids was the hydrolysis of
keratin, like human hair and chicken or duck feathers.

Background

Cysteine, beside methionine the second sulfur containing natural amino
acid, has well established applications in a range of pharmaceuticals, flavors,
bakeries, nutraceuticals and cosmetics. The worldwide consumption of cysteine
in 2003 was about 3700 tons with sales of more than 50 million US dollars.
Roughly 50 % of that market are directed towards pharmaceuticals. The main
portion is used to create active pharmaceutical ingredients, like N-acetyl-cysteine
and S-carboxymethylcysteine, which are used in cough treatment. More than 15
% of the global demand is used as starting material to create reaction or
processed flavors. In addition to that the pet food industry uses close to 20 % of
that volume to create more attractive smelling and tasting food for cats and dogs.

Roughly 200 tons are used worldwide to soften the dough by reducing the disulfide bonds in the wheat proteins. This makes the dough more liquid and the baking process more reproducible. Cosmetic companies in Japan use cysteine as reducing agent for creating permanent waves for the hair. Last but not least cysteines are used more and more in nutraceutical applications, where they act as radical scavenger and anti-aging agents.

Classical production

The major producers of this amino acid are presently Chinese companies. The usual production process is the extraction of the dimer form cystine from keratin hydrolysates, mainly human hair, and the electrochemical reduction to cysteine. As the raw material human hair is not well accepted by the end consumer, especially not by vegetarian people, there is a big demand for alternative production processes. One company succeeded in a synthetic route to produce cysteine, but this route is quite expensive. In addition to that synthetic starting materials are not really the favorites in the food industry. Attempts to establish a fermentation process as common for other amino acids, like glutamate or lysine, have failed so far.

New fermentation process

Wacker-Chemie in Germany developed now a fermentation route for the production of cystine (and cysteine thereof) based on the bacterium Escherichia coli K12, which is very often used in fermentation processes, especially in pharmaceutical applications. This bacterium has the natural ability to produce cysteine from a carbon source and inorganic nitrogen and sulfur sources. Therefore it is not necessary to introduce foreign genes into that organism. Of course the biosynthesis for this energy consuming synthesis has to be highly regulated in nature, otherwise there would be a waste of resources for that bacteria. The main principle of the regulation is the feedback inhibition of key enzymes in that biosynthesis route. Wacker-Chemie did now a surgery type approach to eliminate this feedback inhibitions.

The precursor of cysteine is the amino acid serine. The biosynthesis of this amino acid is mainly regulated by feedback-inhibition. The enzyme , which is sensitive to higher serine concentrations in the cell is the 3-phosphoglycerate dehydrogenase. The serA, coding for this enzyme, was subjected to a mutagenesis procedure to obtain mutants with a reduced feedback inhibition but

132

a sufficient high specific activity. In two subsequent steps serine is activated to O-acetyl serine and sulphur is incorporated to yield cysteine. The biosynthesis of cysteine is again regulated by feedback inhibition. The enzyme sensitive to cysteine is serine transacetylase. The cysE, coding for this enzyme, was again subjected to a mutagenesis procedure and the desired allels coding for feedback resistant enzymes were obtained. Escherichia coli cells, carrying these two types of altered enzymes, produce already cysteine and a secretion to the medium is obtained but the amounts are not sufficient for an industrial production. Because the secretion rate from the cytoplasma to the medium was limited, the third target was to improve the transport from the cytoplasma into the medium. To get this amounts out of the cell across the membrane of this organism, a natural occurring "efflux pump" was for the first time detected and used. The combination of these three optimizations and an adaption of the amount of the natural sulphur regulator CysB resulted in a deregulated organism, which is now synthesizing big amounts of cysteine. The result of that research is an industrial organism (4 patents for the optimization of the biosynthesis, see figure and patent references), which is secreting big amounts of cysteine in the fermentation medium. Due to the large airation of that medium, cysteine is oxidized to cystine, which is precipitating out of the broth, purified and used directly or again electrolytically reduced to cysteine. Due to the fact that the medium consists only of a carbon source (dextrose) and inorganic salts and trace elements, the cystine and the cysteine have vegetarian, Halal, Kosher and natural characteristics.

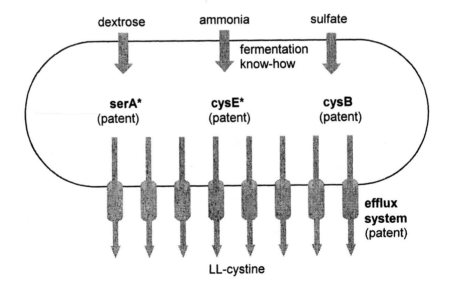

Flavor application

For application in flavors these cysteines are from a chemical point of view identical to the established ones. Therefore they can be exchanged very quickly in existing formulations. They differ only in the origin of the starting material and thereby they can add value to the final flavors. They are fully approved as food ingredient in US and Europe, as they fulfill all the regulations of FCCIV and EC 2000/63. In US the fermentation based cysteines combine uniquely the characteristics of natural origin and vegetarian source and also Kosher and Halal properties in one starting material. The production method by bacteria does not cause a GMO label due to the fact, that there was no gene from another species introduced in that production strain, because the ability to produce cysteine occurs naturally in Escherichia coli, but is also ensured that no bacteria, proteins and DNA are remaining in the product. The production method is now established in industrial scale and supplies the industries already with premium grade cysteines of consistent quality and from unlimited renewable raw materials.

References:
P. Heinrich and W. Leinfelder, WO 97/15673
C. Winterhalter and W. Leinfelder, US 5 972 663
T. Maier and C. Winterhalter, WO 01/27307
G. Wich, W. Leinfelder and K. Backman, US 6 180 373

Process Flavors from Classical Maillard Reactions

Chapter 11

Influence of High Hydrostatic Pressure on Aroma Compound Formation in Thermally Processed Proline–Glucose Mixtures

Peter Schieberle, Thomas Hofmann, and Frank Deters

Chair for Food Chemistry, Department of Chemistry, Technical University of Munich, Lichtenbergstrasse 4, D–85748 Garching, Germany

Application of an Aroma Extract Dilution Analysis on an extract isolated from a glucose/proline mixture thermally processed (90 min, 100°C) at normal pressure (NP) in MPOS buffer confirmed the popcorn-like smelling 2-acetyl-1-pyrroline and 2-acetyltetrahydropyridine as the key odorants formed. However, reacting the same mixture under high hydrostatic pressure (HHP), completely changed the profile of key odorants by minimizing the formation of ACPY, but significantly enhancing the odor intensity of the caramel-like smelling odorants 2-hydroxy-3,4-dimethylpent-2-en-1-one and also 2-hydroxy-3-methyl-pent-2-en-1-one (HMC). Quantitative studies revealed that at HHP much higher amounts of carbohydrate degradation products, such as 2-oxopropanol, are formed. Based on labeling experiments using the CAMOLA technique, a new reaction sequence could be verified for HMC formation at HHP by an Aldol condensation of 2 molecules of 2-oxopropanol followed by elimination of 2 molecules of water.

Introduction

High hydrostatic pressure (HHP) is known to influence the equilibrium of chemical reactions and according to Le Chatelier's law, a reaction is accelerated under pressure, if, e.g., a contraction in the reaction volume occurs (*1*). Other effects influencing reactions under HHP are changes in polarity or the formation of charged intermediates (*2*).

Applications of HHP in food processing were introduced for the first time about 15 years ago (*3*). The basic idea of this technique is to inactivate microorganisms at low temperatures. However, for some foods, such as fruit juices, this "pasteurization" at low temperature offers the opportunity to avoid, e.g., flavor changes by thermal treatments. Although the stability of enzymes and vitamins under HHP has been studied quite often in the past decade (*4*), comparative studies on differences in aroma compounds of foods thermally treated with or without application HHP are scarcely available.

The Maillard reaction between reducing carbohydrates and amines is among the most important flavor generating reactions in thermally processed foods (*5*). Thus, it might be expected that in foods treated with HHP, but at low temperatures, some of the typical aroma compounds might not be formed. Only two studies about the influence of HHP on the formation of volatiles in Maillard model systems are currently available (*6, 7*). Bristow and Isaacs (*6*) reported that at 100°C, the formation of volatiles from xylose/lysine was generally suppressed when HHP was applied. Hill et al. (*7*) confirmed this observation for a glucose/lysine system. However, it has to be pointed out that the samples analyzed were not reacted in a buffered system and, also, the reaction time of the pressure-treated and untreated sample were not identical.

To correlate differences in the aromas of different samples, first, the most-odor active compounds in the samples treated without HHP have to be identified. The subsequent application of a comparative Aroma Extract Dilution Analysis (AEDA) is a very useful tool to screen and compare odor activities of the same odorants in two different samples. To gain first insights into the influence of HHP on aroma formation in Maillard type reactions, the key odorants formed at 100°C in proline/glucose mixtures treated under HHP or at normal pressure were compared. Labeling experiments were then performed to elucidate the influence of HHP on formation of transient intermediates and an pathways leading to selected key odorants.

Material and Methods

Chemicals

3-Hydroxy-4,5-dimethyl-2(5H)-furanone (Sotolon), 4-hydroxy-2,5-dimethyl-3(2H)-furanone (Furaneol) and 2-hydroxy-3-methyl-2-cyclopenten-1-one (Cyclotene) were from Aldrich (Steinheim, Germany). 2-Acetyltetrahydropyridine and 2-acetyl-1-pyrroline were synthesized as previously described (*8*). 2-Hydroxy-3,4-dimethyl-2-cyclopenten-1-one was a gift from Dr. I. Blank (Nestle, Vers-chez-les-blanc, Switzerland).

Reaction system

Glucose (10 mmol) and proline (3.3 mmol) were dissolved in 3-(N-morpholino)propansulfonic acid (MOPS) buffer (50 mL; 0.5 mol/L; pH 7.0) and either reacted at normal pressure (NP) for 90 min at 100°C or pressure-treated at 650 MPa (HHP) for 90 min and 100°C as described recently (*9*). The volatiles formed were isolated by extraction with diethyl ether followed by distillation under high vacuum (*10*). The odor-active compounds were detected by application of the Aroma

Extract Dilution Analysis and subsequently identified using the respective reference compounds for comparison of analytical data. The carbon modul labeling experiments (CAMOLA) were performed as described recently (*11*). α-Dicarbonyl compounds were determined as quinoxaline derivatives as described previously (*12*).

Results and Discussion

In a first experiment, two solutions of glucose and proline were separately reacted in MOPS buffer at a constant pH of 7.0 for 90 min at 100°C. One mixture was treated at HHD, the other at normal pressure (NP). The MOPS buffer was used, because it is known to be stable under HHP and, thus, was able to keep the pH constant, which is an important prerequisite in studies at HHP.

After cooling, the overall aromas of both solutions were evaluated by a sensory panel consisting of 10 members. As shown in Figure 1, the mixture processed under NP exhibited a very intense popcorn-like, roasty aroma. However, in the mixture treated with HHP, this aroma note was quite weak (Figure 1), whereas a caramel-like odor quality was much more pronounced. To elucidate the compounds showing the highest odor activities in both solutions, the volatiles were isolated, separated by GC and evaluated by GC/odorport evaluation. Their odor intensities were then ranked by application of the AEDA.

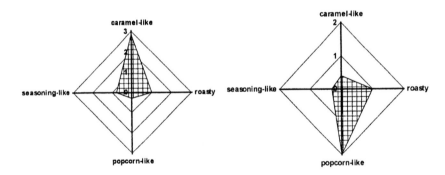

Figure 1. Aroma profiles of thermally processed glucose/L-proline mixtures. Right: normal pressure (NP); left: high hydrostatic pressure (HHP)

In the extract isolated from the mixture reacted at NP, the popcorn-like smelling aroma compounds 2-acetyl-1-pyrroline and 2-acetyltetrahydropyridine showed by far the highest flavor dilution factors (Figure 2). This result is in good agreement with data previously reported on key odorants formed from glucose/proline mixtures reacted in the absence of water (*13*) or in a phosphate buffer, respectively (*14*). In the glucose/proline mixture reacted under HHP, the spectrum of odor-active compounds was completely changed (Figure 3). The popcorn-like smelling 2-acetyltetra-hydropyridine was sensorially not detectable and 2-acetyl-1-pyrroline only showed a low FD-factor. By contrast, the intensely caramel-like smelling 2-hydroxy-3,4-dimethylpent-2-en-1-one followed by the caramel-like smelling 4-hydroxy-2,5-dimethyl-3(2H)-furanone and 3-hydroxy-4,5-dimethyl-2(5H)-furanone, with a seasoning-like odor quality, were the most intense (Figure 3). These data were in full agreement with the results on the overall sensory evaluation showing a very intense caramel-like aroma in the sample treated at HHD (cf. Figure 1).

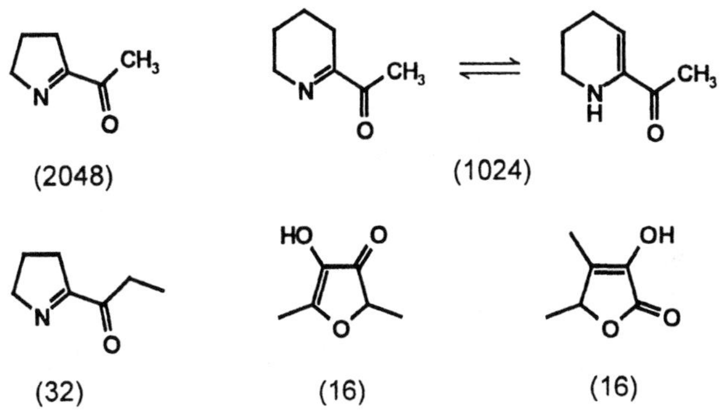

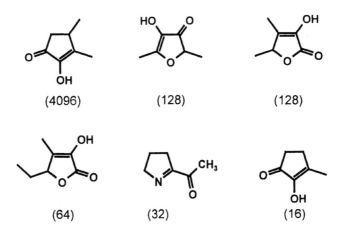

Figure 2. Structures of the most odor-active compounds (FD-factor in parentheses) formed from proline/glucose at 100 °C (90 min) under normal pressure

Figure 3. Structures of the most odor-active compounds formed from glucose/proline at 100 °C (90 min) under high hydrostatic pressure (FD-factor in parentheses)

The results suggested that either 2-acetyl-1-pyrroline and 2-acetyltetrahydro-pyridine are unstable under HHP or, alternatively, that the furanones and cyclopentenones are formed to a higher extent by an enhanced carbohydrate degradation when HHP is applied. To gain some insight into the stability of several aroma compounds during heating under HHP, these were dissolved in MOPS buffer and reacted for 15 min at 100°C either at 650 MPa or at 0.1 MPa. Interestingly, 2-acetyl-1-pyrroline was stable under these conditions at NP, but decreased by about 22 %, when HHP was applied (Table I). However, 4-hydroxy-2,5-dimethyl-3(2H)-furanone, which was stable at NP, was nearly completely degraded after the HHP treatment. By contrast, the cyclopentenone was not affected, when the reaction was performed at 650 MPa. These results, in particular the stability of the cyclopentenone, which was only detected with a low FD-factor in the sample treated at NP (Figure 2), implied that the pressure treatment influenced the reaction cascade leading to the formation of the aroma compounds. Because cyclopentenones can only be formed after cleavage of the carbohydrate skeleton, the pressure treatment obviously led to an enhanced formation of carbohydrate degradation products, such as α-dicarbonyls.

Table I. Stability of selected Maillard aroma compounds[a]

Exp.	Aroma compound	Pressure (MPa)	Amount left (%)
1	2-Acetyl-1-pyrroline	0.1	100
2	2-Acetyl-1-pyrroline	650	78
3	4-Hydroxy-2,5-dimethyl-3(2H)-furanone	0.1	100
4	4-Hydroxy-2,5-dimethyl-3(2H)-furanone	650	10
5	2-Hydroxy-3,4-dimethyl-pent-2-en-1-one	0.1	100
6	2-Hydroxy-3,4-dimethyl-pent-2-en-1-one	650	99

[a] The aroma compound (10 µg) was dissolved in MOPS-buffer (30 mL; pH 7.0; 0.25 mol/L) and heated for 15 min at 100°C.

To check this assumption several α-dicarbonyls and α-oxo-hydroxy compounds, known as degradation products of carbohydrates, were quantified using derivatization with o-phenylene diamine or ethyl hydroxyamine, respectively, as shown in Figure 4. Quantitation was performed by a stable isotope dilution assay using $[^{13}C_4]$-2,3-butandione as the internal standard (9). As an example, the time courses of the formation of 2-oxopropanol in the glucose/proline mixture at HHP or NP are contrasted. In the pressure treated sample, the amount of 2-oxopropanol reached quite high concentrations and a maximum after 60 min at 100°C. The amount of 14800 mg/kg glucose formed was by a factor of about 900 higher as compared to the amounts formed under NP (Figure 5). Similar results were obtained for several other carbohydrate degradation products (9).

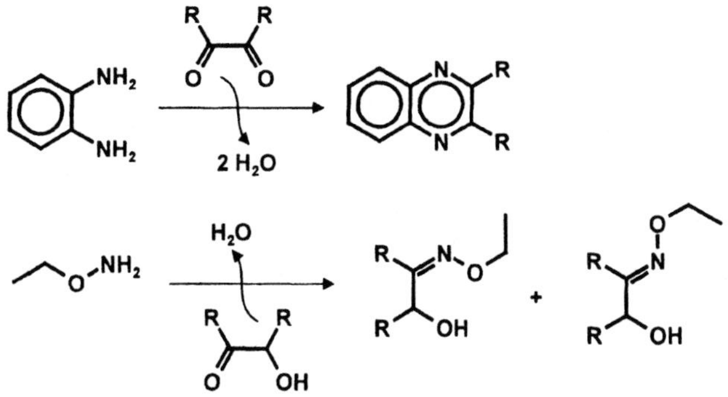

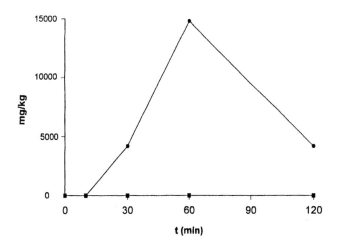

Figure 4. Methods for quantitation of carbohydrate degradation products using derivatization with o-phenylene diamine or ethyl hydroxamine

Figure 5. mounts of 2-oxopropanol formed during thermal processing (100°C, pH 7) of a glucose/proline mixture
(●: high hydrostatic pressure; ■: normal pressure)

Based on these results, it might be concluded that, because of the high concentrations of carbohydrate degradation products in the HHP treated system, retro-Aldol reactions are favored, which could then lead to the formation of higher amounts of, in particular, the cyclopentenones.

In Figure 6 a hypothetical pathway leading to 2-hydroxy-3-methyl-cyclopent-2-en-1-one (HMC) by a retro-Aldol condensation of 2 molecules of 2-oxopropanol or its tautomer, 2-hydroxypropanal, respectively, is shown. To prove that two carbon-3 compounds are in fact involved in the formation of HMC, a 1+1 mixture of $[^{13}C_6]$-glucose and $[^{12}C_6]$-glucose was reacted in the presence of proline and the ratio of isotopomers in the HMC formed were quantified by mass chromatography. Based on this idea, which was recently published by us as "carbon modul labeling (CAMOLA) technique" (11), the "isotope scrambling" of fragments formed as transient intermediates from labeled and unlabeled glucose will indicate, which "carbon moduls", e.g. carbon-2 or carbon-3 fragments, are involved in the formation of the target compounds. In Figure 7, the mass spectrum (MS/CI) of HMC formed under HHP from a 1+1 mixture of labeled and unlabeled glucose is displayed. The isotopomer at m/z 116 corresponds to the triply labeled HMC. For statistical reasons, the abundance of this isotopomer has to be doubled as compared to the isotopomers at m/z 113 and m/z 119, if a C-3 compound is the only intermediate involved. Because this was clearly reflected in the intensities of the respective ions (an exact calculation of the relative abundances of each mass was done by the computer of the mass spectrometer), these data corroborate the reaction pathway shown in Figure 6.

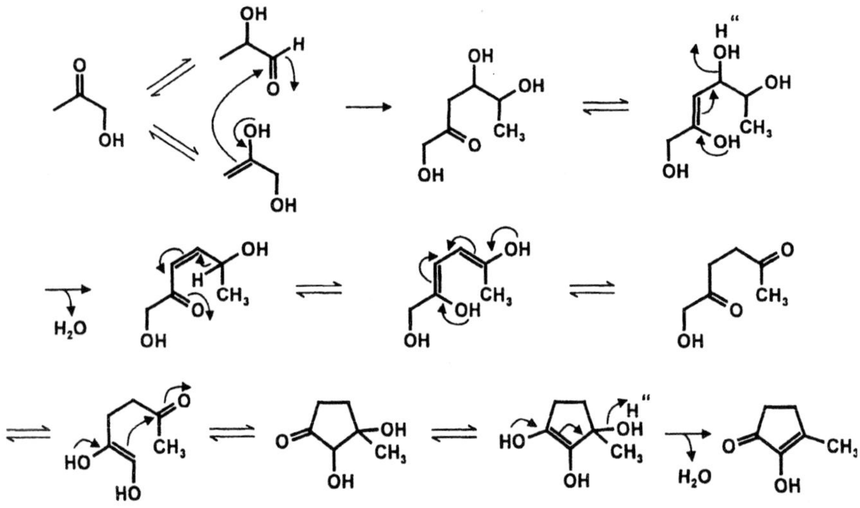

Figure 6. Hypothetical pathways leading to the formation of 2-hydroxy-3-methyl-cyclopent-2-en-1-one from 2-hydroxypropanal and 2-oxopropanol by an Aldol reaction

It might be speculated that a certain portion of the 2-oxopropanol may be formed by a condensation of a C-2 and a C-1 fragment (e.g. acetaldehyde and formaldehyde) prior to the formation of the HMC. Because mass spectrometry cannot distinguish between the different possible positions of carbon 13 in the isotopomers, this possibility was not ruled out by the above experiments.

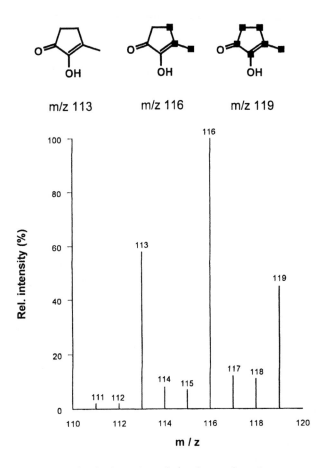

m/z 113   m/z 116   m/z 119

Figure 7. Isotopomers of 2-hydroxy-3-methylcyclopent-2-en-1-one generated under high hydrostatic pressure from $^{12}C_6$- and $^{13}C_6$-glucose in the presence of proline

To indicate that 2-oxopropanol is really formed as an intact C-3 carbon modul from glucose prior to forming HMC as shown in Figure 6, three different mixtures were reacted under HHP:

A: [$^{12}C_6$]-glucose/proline, B: [$^{13}C_6$]-glucose/proline and C: [$^{12}C_6$]- and [$^{13}C_6$]-glucose/praline (1+1). The 2-oxopropanol formed from the three mixtures was derivatized with ethyl hydroxamine and the derivatives were then analyzed by MS/CI.

144

Based on the data obtained (Figure 8) from the control experiments A and B, which only showed the unlabeled (m/z 118) or the fully labeled 2-oxopropanol (m/z 121), the two isotopomers obtained from mixture C clearly indicated that no isotope scrambling had occurred (the fragments at m/z 119 and m/z 120 are caused by M+1 or M-1 fragments of m/z 118 or m/z 121, respectively). Thus, the data confirmed that the 2-oxopropanol present in the reaction system is exclusively formed as a carbon-3 modul from either glucose or the carbon-13 labeled glucose, respectively.

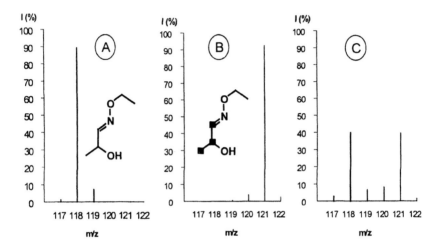

Figure 8. Isotopomers of 2-hydroxypropanal (as ethoxime derivatives) formed from $^{12}C_6$-glucose/proline (A); $^{13}C_6$-glucose/proline (B) and a 1+1 mixture of $^{12}C_6$-/$^{13}C_6$-glucose/proline (C) racted under high hydrostatic pressure

In summary, the results indicated that the Maillard reaction of proline/glucose is significantly influenced by high hydrostatic pressure. Literature data suggesting a reduced formation of aroma volatiles under HHP were not verified. However, the results indicated HHP as an additional parameter to influence the structure of the aroma compounds formed in Maillard-type reactions. Because obviously the carbohydrate fragmentation is enhanced at HHP, aroma compounds being formed by a retro-Aldol reaction of such transient intermediates predominate, whereas N-containing odorants are reduced. Because the instability of N-compounds, such as 2-acetyl-1-pyrroline did not account for their reduced occurrence under HHP, obviously the high amounts of α-dicarbonyls generated may inhibit their formation to a certain extent.

References

1. Klärner, F.G.; Wurche, F. *Journal fuer praktische Chemie*, 2000, *342*, 103-162.
2. LeNoble, W.J. The influence of pressure on chemical reactions in solution. *Chemie in unserer Zeit*, 1983, *17*, 162-162.
3. Knorr, D. *Food Technology*, 1993, *47*, 609-636.
4. Pfister, M.K.H.; Dehne, L.I. *DLR*, 2001, *97*, 257-264.
5. Schieberle, P.; Hofmann, T. In: Advances in Flavours and Fragrances. From the Sensation to the Synthesis (Swift K.A.D., ed.) The Royal Society of Chemistry, Cambridge, 2002, pp. 163-177.
6. Bristow, M.; Isaacs, N.S. *J. Chem. Soc. Perkin Trans.* 1999, *2*, 2213-2218.
7. Hill, V.M.; Isaacs, N.S.; Ledward, D.A.; Ames, J.M. *J. Agric. Food Chem.* 1999, *44*, 594-598.
8. Schieberle, P. *Z Lebensm. Unters. Forsch.* 1990, *191*, 206-209.
9. Deters. F.; Schieberle, P. *J. Agric. Food Chem.*, 2004, submitted.
10. Engel, W.; Bahr, W.; Schieberle, P. *Eur. Food Res. Technol.* 1999, *209*, 237-241.
11. Schieberle, P.; Fischer, R.; Hofmann, T. In: Flavour Research at the Dawn of the Twenty-first Century, Proceedings of the 10th Weurman Flavour Research Symposium (Le Quéré J.-L., Étiévant J.X., eds.) 24.-28.06.2002, Beaune, London, 2003, pp. 447-452.
12. Hofmann, T. *Eur. Food Res. Technol.* 2003, *209*, 113-121.
13. Roberts, D.D.; Acree, T.E. In: ACS Symposium Series, Thermally generated flavors, 1994, pp. 71-79.
14. Hofmann, T.; Schieberle, P. *J. Agric. Food Chem.* 1998, *46*, 2721-2726.

Chapter 12

Effect of Acetate Concentration on Maillard Browning in Glucose–Glycine Model Systems

Cathy Davies

Department of Animal and Food Sciences, University of Delaware, Newark, DE 19716

This study investigated the effect of acetate on the rate of Maillard browning in glucose-glycine model systems. Solutions, containing glucose, glycine and acetate (0-0.5 M, pH 5.5), were incubated (50°C) and A_{420} was measured over time. Data were analyzed by linear regression ($A = k_0t + c$) and by non-linear regression ($A = at^2 + bt$), where A is A_{420}; t, time (h); k_0 rate (h^{-1}); c intercept; a and b are calculated parameters. For buffered solutions, A_{420} increased with time and with increasing acetate, glucose or glycine concentration. k_0 increased from ($1.24 \forall 0.07$ to $7.74 \forall 0.54$) x 10^{-3} h^{-1} with increasing acetate concentration from 0.1 to 0.5 M. Regression coefficients ranged from 0.8485 to 0.9860. From the polynomial equation, parameters a and b increased from ($2.84 \forall 0.035$ to $24.7 \forall 0.88$) x 10^{-6} and ($-2.0 \forall 1.0$ to $-10.0 \forall 2.0$) x 10^{-4}, respectively, as acetate was increased. These results suggest that acetate participates directly in the Maillard reaction.

Introduction

The Maillard reaction has been extensively studied in buffered solutions (*1; 2; 3; 4; 5; 6; 7; 8; 9*). The Maillard reaction is inhibited at low pH and accelerated at high pH. The rate of browning is increased by the presence of buffer. Common buffers used are citrate, acetate and phosphate. Earlier research on the effect of different buffer salts or acids, suggests that these compounds accelerate the browning reaction of amino acids with glucose (*10; 11; 12*) and fructose (*13*). The addition of organic acids such as citric, tartaric and succinic acid to model systems investigating the browning of ascorbic acid increased the rate of browning with increasing citric acid (*14*). Ascorbic acid was more stable in orange juice than in phosphate buffered solutions at the same pH (*15*). The presence of organic acids such as tartaric or succinic acid also increased the rate of ascorbic acid degradation in these systems (*14*). Thus, the effect of organic acids and buffer salts is of considerable practical significance as fruit and vegetables, in particular, contain sugars, amino acids and both phosphates and organic acid salts. In addition, the use of different buffers may alter the flavors formed when developing processed flavors.

The mode of action of buffers on the Maillard reaction has been proposed to be (*11*):

1) Under mildly alkaline conditions, buffers reduce the extend of the decrease in pH caused by acidic reaction products, thus maintaining a higher reaction rate

2) In acid solutions, the buffer takes part in general acid-base catalysis. The Amadori rearrangement of glucose amines to fructosamines and their subsequent decomposition are subject to general acid-base catalysis (*16*).

Phosphate buffers, in particular, are thought to catalyze the Maillard reaction. Increasing phosphate concentration in a glucose-glycine model system (pH 7; 25 °C) increased the pseudo zero order rate constant for browning (*12*). The reason suggested for the effect of phosphate was the catalytic effect of phosphate anions on the Amadori rearrangement (*10; 12; 17*). In addition to acting as a general base catalyst, phosphate increased the polarographic wave height of aldose sugars (*10*), suggesting that phosphate increases the rate of sugar mutarotation. The physical and chemical properties of reducing sugars are affected by the relative concentration of their different tautomeric forms. Specifically, the concentration of the open chain forms greatly influences their reactivity towards the Maillard reaction (*18*). For example, solutions of phosphorylated sugars, such as glucose-6-phosphate and ribose-5-phosphate, are more reactive, and have a higher concentration of open chain sugars, than their non-phosphorylated counterparts (*19*) (*20*). Phosphate ions may also catalyze the decarboxylation of amino acids (*20*). However, doubling the phosphate concentration reduced the formation of advanced glycation end products of bovine serum albumin, despite the higher buffer concentrations stabilizing the pH (*21*).

The effect of acetate on loss of sulphite from a fructose-glycine-S(IV) model system was investigated. The rate of the reaction, as estimated by the loss of S(IV), increased with increasing acetate concentration. In addition, a glycine independent reaction was observed (*13*).

The objective of this work was to investigate the effect of acetate buffer on the browning glucose-glycine in model systems.

Methods

Model systems containing glucose (0.0 - 0.4 M), glycine (0.0 - 0.4 M) and acetate buffer (0 - 0.5 M, pH 5.5 adjusted with glacial acetic acid) were incubated at 50 °C and absorbance (A_{420}) was measured at timed intervals using a Beckman DU60 spectrophotometer with water as a blank. Readings were taken until absorbance reached 2.0. The pH was measured with a Corning 443i pH meter. All reaction mixtures were prepared in triplicate.

Data was analyzed in two ways:

(1) Linear regression was determined for the later stages of browning using $A = k_0 t + c$.

(2) Non-linear regression for all of the data using the equation, $A = at^2 - bt$, where A is absorbance (420 nm), t = time (h), k_0 = gradient, c = intercept, a and b are calculated parameters by non-linear regression.

Linear analysis was carried out using Excel and the non linear regression analysis was carried out with NLREG (Version 5.2 Phillip H. Sherrod ©1991-2001).

Results and Discussion

A_{420} increased with time in all solutions, except for the systems with either no acetate or no glucose. The absorbance-time plot showed that change in absorbance with time was not linear as the rate of browning increased with time (Figure 1). This is the typical shape of the reaction curve, and the induction period and shape of the curve have been explained as the initial absence of a reactant which is subsequently formed during heating (*22*).

A slight fall in pH from 5.5 to 5.3, was observed over time for all buffered solutions. In the model system with no buffer, pH was not adjusted; initial pH was 5.9 and final pH was 4.9.

Absorbance increased with increasing acetate concentration (Figure 2). When no buffer was present, however, the change in absorbance was so small in the time of the experiment (0.03 in >300 h) that essentially no reaction took place.

The kinetics of the Maillard reaction have been determined using different models. The Maillard reaction has been simplified to fit either a zero or first order reaction despite that neither are a good model of browning due to the complexity with many side and parallel reactions. The simplest method for determining the kinetics of Maillard browning is to determine the rate of the reaction from the gradient of the linear portion of the browning curve after the induction phase is over (Figure 1) (*23*). This is known as the pseudo zero order rate constant (k'_0) and gives the rate of the late stage of Maillard browning.

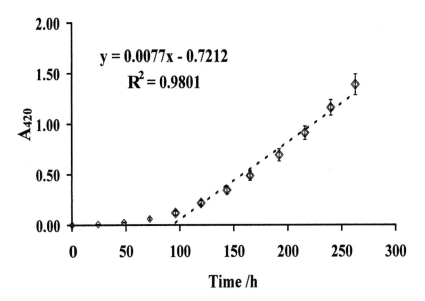

$$y = 0.0077x - 0.7212$$
$$R^2 = 0.9801$$

Figure 1: Absorbance (A_{420}) with time for glucose (0.2 M) - glycine (0.2 M) in acetate (0.5 M) pH_i 5.5; 50 °C; (symbols are experimental data, line is calculated regression line).

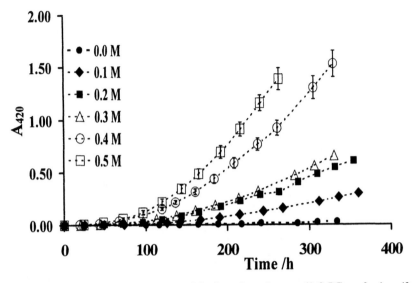

Figure 2: Change in absorbance with time for glucose (0.2 M) - glycine (0.2 M) in acetate (0 - 0.5 M) pH$_i$ 5.5, 50 °C (experimental data shown only).

Pseudo zero order rate constants were calculated for all model systems containing buffer (Table 1). The rate of browning increased with increasing acetate concentration. Such a linear equation, however, assumes that the later stages of browning can be approximated to a zero order reaction. Zero order assumes that browning is not affected by either reactant concentration or by the earlier stages of the Maillard reaction. Earlier research suggested that not only is Maillard browning reactant concentration dependent but the reaction before browning occurs has the greatest influence on browning. Thus, a zero order model of the data ignores these important factors.

Table 1: Pseudo zero order rates and intercepts for browning (A$_{420}$) of glucose (0.2 M) - glycine (0.2 M) in acetate (0.1 - 0.5 M) pH$_i$ 5.5; 50 °C.

Acetate (M)	$10^3 \times k_0 /h^{-1}$	Intercept (c)	r^2
0.1	1.24 ∀ 0.07	-0.162 ∀ 0.01	0.9860 ∀ 0.0022
0.2	2.26 ∀ 0.2	-0.241 ∀ 0.03	0.9744 ∀ 0.0109
0.3	2.73 ∀ 0.2	-0.300 ∀ 0.03	0.9787 ∀ 0.0004
0.4	6.44 ∀ 0.5	-0.709 ∀ 0.04	0.9815 ∀ 0.0029
0.5	7.72 ∀ 0.5	-0.721 ∀ 0.06	0.9798 ∀ 0.0062

An exponential equation was used to fit the data to test for first order behavior. While there was a good fit to an exponential equation, the value of the parameters determined could not be used to predict the effect of acetate on browning.

It is not surprising that simple kinetics do not model browning accurately, since measuring browning is determining the concentration of a product after a series of reactions. Polynomial equation have been used to characterize these absorbance-time relationships. These are developed from idea that a reactant is produced during the reaction and are based on a consecutive 3-step mechanism (1):

$$A + S \xrightarrow{k_a} I_1 \xrightarrow{k_b} I_2 \xrightarrow[+A]{k_c} B$$

Where S, A and B denote the reducing sugar, amino acid and product responsible for browning (melanoidins), respectively. I_1 and I_2 are intermediates and k_a, k_b, and k_c are rate constants. Assuming that there is no rate limiting step after the formation of I_2, and I_2 and I_1 are dependent on the initial reaction between sugar and amino acid, the kinetic for each equation can be hypothesized to be:

$$\frac{d[I_1]}{dt} = k_a[A][S]$$

$$\frac{d[I_2]}{dt} = k_b[I_1]$$

$$\frac{dB}{dt} = k_c[A][I_2]$$

From integration, the amount of browning present is given by:

$$B = k'[A]^2[S]t^2$$

where k' represents a combination of k_a, k_b, and k_c (1). To model browning a second order polynomial equation, $A_{420} = at^2 + bt$, was developed. Parameters a and b were determined using non linear regression analysis. The data showed that the absolute values of these parameters increased with increasing acetate concentration (Table 2).

Table 2: Values of parameters a and b for browning (A_{420}) of glucose (0.2 M) - glycine (0.2 M) in acetate (0.1 M - 0.5 M), pH$_i$ 5.5, 50 °C

Acetate (M)	10^6 x a	10^4 x b ·	r^2
0.1	2.84 ± 0.0333	-2.0 ± 0.096	0.9990
0.2	5.41 ± 0.357	-2.0 ± 1.020	0.9699
0.3	6.88 ± 0.309	-3.0 ± 0.825	0.9870
0.4	15.7 ± 0.599	-5.0 ± 1.580	0.9911
0.5	24.7 ± 0.0875	-10.0 ± 1.860	0.9915

The change in parameters a and b with acetate concentration was non linear and could be further analyzed using nonlinear regression to model the relationship between acetate concentration, time and absorbance. Parameter a was found to have the best fit with a second order polynomial:

$$a = m[Ac] + n[Ac]^2$$

Parameter b had a best fit with an exponential equation:

$$b = pe^{-q[Ac]}$$

Where m, n, p and q are calculated parameters and [Ac] is acetate concentration. If these relationships are placed into equation 1, the relationship of absorbance with time and acetate concentration is:

$$A_{420} = (pe^{-q[Ac]})t + (m[Ac] + n[Ac]^2)t^2$$

For browning of glucose and glycine (both 0.2 M, 50 °C, pH 5.5) the values for the calculated parameters are shown in Table 3. Presumably these values will change with different reactant concentrations, reactants, buffers, and conditions.

Table 3: Values of non-linear regression parameters for browning in glucose (0.2 M) - glycine (0.2 M), with acetate, pH 5.5, 50 °C. Browning was measured as A_{420}

parameter	value
m	$(8.99 \ \forall\ 1.98) \times 10^-$
n	$(3.23 \ \forall\ 8.35) \times 10^-$
p	$(5.64 \ \forall\ 1.78) \times 10^-$
q	$-5.71 \ \forall\ 0.69$

Browning increased with increasing glucose or glycine concentration. Increasing the concentration of glucose or glycine increased the absolute values of a and b, regardless of whether the concentration of either glucose and glycine was increased (Table 4). This was surprising since it was thought that the amino acid played a more important role than the reducing sugar owing to its involvement in more than one reaction step.

Table 4: Parameters a and b for different glucose, glycine, and acetate concentrations

10^4 x Parameter b		Acetate (M)		
Glucose (M)	Glycine (M)	0.5	0.3	0.1
0.2	0.4	-32.1	-13.6	6.0
0.2	0.2	-10.0	-2.8	-1.5
0.4	0.2	-49.3	-8.1	2.3

10^4 x Parameter a		Acetate (M)		
Glucose (M)	Glycine (M)	0.5	0.3	0.1
0.2	0.4	0.85	0.27	0.04
0.2	0.2	0.24	0.07	0.03
0.4	0.2	1.04	0.22	0.05

In the absence of glycine (Figure 3), solutions containing glucose turned brown when acetate was the buffer. This was not seen when citrate was the buffer. A glycine-independent reaction was also observed for fructose in acetate buffer (pH 4.5 or 5.5/ 55 °C) (13). This observation suggests that acetate somehow catalyzes sugar degradation at temperatures lower than expected. The

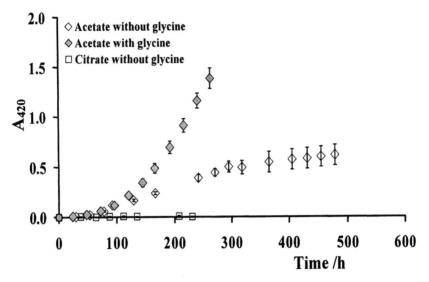

Figure 3: Browning (A$_{420}$) of glucose (0.2 M) - acetate (0.5 M) or citrate (0.5 M) with and without glycine (0, 0.2 M), pH$_i$ 5.5; 50 °C.

mechanism of acetate catalysis is not understood and requires further investigation.

Conclusion

When the acetate ion concentration was increased from 0.1 to 0.5 M in glucose-glycine model systems, the rate of browning (A$_{420}$) increased six times and absolute values of parameters a and b increased approximately nine and five times respectively. Browning was also observed when glycine was absent. This change is not caused by improved pH control, but by another effect of acetate suggesting that acetate is as important as glucose and glycine in the browning reaction mechanism.

Parameters a and b were obtained from non linear regression of the browning data using the equation $A_{420} = at^2 + bt$. Increasing the glucose or glycine concentration increased the absolute values of a and b; when the acetate concentration was 0.3 or 0.5 M and the concentration of either glucose or glycine was doubled, the value of a increased three fold.

References

1. Davies, C. G. A.; Wedzicha, B. L.; Gillard, C. *Food Chemistry* **1997**, *60*, 323-329.
2. Bell, L. N.; White, K. L.; Chen, Y. H. *Journal of Food Science* **1998**, *63*, 785-788.
3. Lievonen, S. M.; Roos, Y. H. *Journal of Food Science* **2002**, *67*, 2100-2106.
4. Davidek, T.; Clety, N.; Aubin, S.; Blank, I. *Journal of Agricultural and Food Chemistry* **2002**, *50*, 5472-5479.
5. Moreno, F. J.; Molina, E.; Olano, A.; Lopez-Fandino, R. *Journal of Agricultural and Food Chemistry* **2003**, *51*, 394-400.
6. Thornalley, P. J.; Langborg, A.; Minhas, H. S. *Biochemical Journal* **1999**, *344*, 109-116.
7. Hill, V. M.; Ledward, D. A.; Ames, J. A. *J. Agric. Food Chem.* **1996**, *44*, 594-598.
8. Zyzak, D. V.; Richardson, J. M.; Thorpe, S. R.; Baynes, J. W. *Archives of Biochemistry and Biophysics* **1995**, *316*, 547-554.
9. Haleva-Toledo, E.; Naim, M.; Zehavi, U.; Rouseff, R. L. *Journal of Agricultural and Food Chemistry* **1997**, *45*, 1314-1319.
10. Burton, H. S.; McWeeny, D. J. *Nature* **1963**, *197*, 266-&.
11. Saunders, J.; Jervis, F. *Journal of the Science of Food and Agriculture* **1966**, *17*, 245-&.
12. Bell, L. N. *Food Chemistry* **1997**, *59*, 143-147.
13. Swales, S. E.; Wedzicha, B. L. *Food Chemistry* **1995**, *52*, 223-226.
14. Clegg, K. M. *Journal of the Science of Food and Agriculture* **1966**, *17*, 546-549.
15. Horton, P. B.; Dickman, S. R. *Journal of Food Protection* **1977**, *40*, 584-585.
16. Reynolds, T. M. *Australian Journal of Chemistry* **1959**, *12*, 265-274.
17. Potman, R. P.; Van Wijk, T. A. In *Thermal Generation of Aromas*; T. H. Parliament; R. J. McGorrin and C.-T. Ho, Eds.; American Chemical Society: Washington, **1989**; pp 182-195.
18. Aaslyng, M. D.; Larsen, L. M.; Nielsen, P. M. *Zeitschrift Fur Lebensmittel Untersuchung und Forschung A: Food Research and Technology* **1999**, *208*, 355-361.
19. Yaylayan, V. A.; Ismail, A. A. *Carbohydrate Research* **1995**, *276*, 253-265.

20. Yaylayan, V. A.; Machiels, D.; Istasse, L. *Journal of Agricultural and Food Chemistry* **2003**, *51*, 3358-3366.
21. Tessier, F.; Birlouez-Aragon, I. *Glycoconjugate Journal* **1998**, *15*, 571-574.
22. Haugaard, G.; Tumerman, L.; Siverstri, H. *Journal of the American Chemical Society* **1951**, *73*, 4594-4600.
23. Labuza, T. P. *Journal of Chemical Education* **1984**, *61*, 348-358.

Chapter 13

Role of Phosphate and Carboxylate Ions in Maillard Browning

George P. Rizzi[1,2]

[1]Proctor and Gamble Company, Winton Hill Business Center,
Cincinnati, OH 45224
[2]Current address: 542 Blossomhill Lane, Cincinnati, OH 45224–1406

Phosphate and other polyatomic anions will accelerate the rate of Maillard browning. In this work we measured the effect of phosphate and carboxylate ions on the rates of visual color formation for a series of reducing sugars with or without the addition of β-alanine. By using an inert buffer system we were able to estimate individual contributions of sugars and the amino acid to color formation. Significant browning was observed for sugars alone suggesting that polyatomic anions contribute to Maillard browning by providing reactive intermediates directly from sugars. A mechanism is proposed for decomposition of sugars by polyatomic anions and efforts to trap reactive species as quinoxaline derivatives using o-phenylenediamine (OPD) are described. The results of this study point out how complications may arise from the popular usage of phosphate buffers in the study of Maillard reactions.

Introduction

The Maillard reaction is a long-recognized source of flavor and color in processed foods. Today although much has been learned about Maillard chemistry the effective control of the reaction during food processing remains elusive. In a recent review Martins et al. (1) summarized Maillard chemistry and pointed out how multiresponse kinetic modeling might be applied to control food quality attributes resulting from the reaction. A possible addition to improve this promising approach is to incude a parameter relating to the effects of catalytic agents on browning. Toward this end we wish to report our observations on the catalytic role of phosphate and carboxylate ions in Maillard browning.

The accelerating effect of phosphate on Maillard browning is well known and was first studied in detail by Reynolds (2). Reynolds observed that no inorganic phosphate was consumed during the reaction and that no stable intermediate containing covalently bound phosphorus could be detected. A more extensive mechanistic study of phosphate catalysis based on kinetic data was reported by Potman and van Wijk in 1989 (3). From a pH study it was concluded that the catalytic species is the dihydrogen phosphate ion which was presumed to act as a base for catalyzing the Amadori rearrangement. In concert with this argument phosphate was recently shown to accelerate the degradation of an Amadori compound into an amino acid and the parent sugar (4). However, in spite of these results the exact mechanism for phosphate involvement is still not clear. For example, Bell (5) had later concluded that phosphate acts as a bifunctional (acid/base) catalyst to speed up the glycosylation of amino acids.

The independent reaction of sugars with phosphate has not received much attention as a possible contributor to Maillard browning. Potman and van Wijk (3) observed that the rate of glucose loss in a glycine/glucose reaction was dependent on phosphate concentration, but the effect of this phenomenon on browning was not considered. In a related mechanistic study, Weenen and Apeldoorn (6) observed low yields of sugar cleavage products, glyoxal and pyruvaldehyde in sugar/phosphate reactions without amino acids.

Experimental Section

General Reaction Procedure

Maillard browning reactions were usually performed by refluxing 0.10 M aqueous phosphate, acetate, succinate or 2,2-bis(hydroxymethyl)-2,2',2"-nitrilotriethanol ("bis-tris") buffer solutions containing various carbohydrates (0.10 M) and β-alanine (0.033 M) for a periods ranging up to 300 min. at approximately neutral pH. At increasing time intervals small samples of hot reaction mixtures were withdrawn with pre-chilled syringes and immediately diluted 1:4 with 23°C water prior to absorbance measurement (vs. pure water) in 1 cm cuvettes at 420 nm.

Preparation and Analysis of Quinoxaline Derivatives

Following an 80 min reaction, ribose/phosphate buffer mixtures were rapidly cooled to 25° C, treated with o-phenylenediamine and incubated at 50°C for 30 min. Quinoxalines and unreacted OPD were extracted and concentrated for GC/MS analysis or separation by preparative TLC. Volatile quinoxaline and methylquinoxalines were identified by R_t and MS of available standards. 2-Ethylquinoxaline was tentatively identified by its MS alone. Non-volatile quinoxalines were isolated by preparative TLC and tentatively identified through literature R_f , NMR and UV data. Addition of OPD at the start of ribose/phosphate buffer reactions produced somewhat higher yields of the same quinoxalines in a similar ratio. No evidence for quinoxaline products was found in control experiments run in bis-tris buffer.

Results and Discussion

Effects of Various Buffers on Browning

Maillard browning was followed by measuring changes in absorbance (A_{420}) versus time in 0.10 M buffer solutions at reflux temperature (ca. 100° C) initially containing reducing sugars (0.10 M) and β-alanine (0.033 M). β-Alanine was chosen as the common amino acid to minimize complications resulting from competitive Strecker degradations. Reactions were generally monitored during 160 min when visual color ranged from colorless to shades of yellow to orange and finally to red. Reaction pH initially at 7.3 ± 0.04 decreased slightly during most reactions indicating the formation of strong acid(s). However, the fact that large A_{420} changes were consistently induced by phosphate at nearly constant pH suggested that phosphate was acting independently of hydrogen and/or hydroxide ion concentration to influence browning. Absorbance versus time roughly followed first order kinetics and data are typified by the reactions of xylose shown in Figure 1.   Various reducing sugars differed in their propensity for browning in contact with β-alanine (Figure 2). The effect of phosphate was estimated by comparing the browning produced in phosphate buffer with the browning obtained in similar reactions using a relatively unreactive,"bis-tris" buffer (pKa 6.5). In Figure 2 absorbances are shown at a single reaction time (80 min) for comparison. Replicate experiments typically exhibited an absorbance precision ranging from

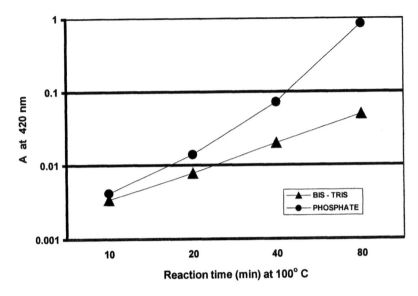

Figure 1. Xylose/β-alanine browning at 100°C and pH 7.3 in phosphate and bis-tris buffers

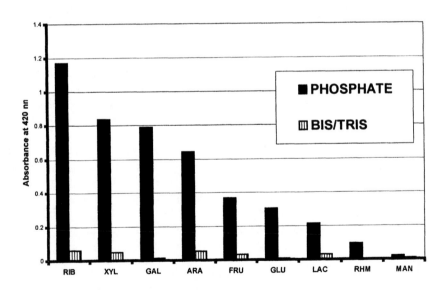

Figure 2. Browning of sugars with β-alanine after 100°C/80 min. in phosphate or bis-tris buffer at pH 7.3

± 3% to ± 20% with the lowest absorbances showing the greatest error. All sugars showed enhanced Maillard browning in phosphate buffer versus bis-tris buffer. In general, pentoses (ribose, xylose and arabinose) displayed greater browning than hexoses (glucose, mannose and rhamnose). Among the hexoses galactose and fructose showed relatively high browning potential comparable with the pentoses. Browning tendencies among individual

sugars are not clearly understood, but are probably related to differing amounts of acyclic (carbonyl-form) isomers available at equilibrium in buffers at 100°C (7).

Enhanced browning was also observed in carboxylate-based buffers. The A_{420} produced by ribose/β-alanine in 0.10 M acetate buffer under Figure 2 conditions (0.15 after 80 min/100° C) was significantly less than 1.17 seen in phosphate, however it was still greater than 0.061 obtained in inert bis/tris buffer. The pKa of acetic acid (4.76) was too low to enable effective buffer action near neutral pH, and a relatively large pH change during the reaction (- 1.33) may have influenced color formation in acetate. Dibasic organic acids with pK_2 closer to the pK_2 of phosphoric acid (7.12) were better buffers near pH 7 and also produced enhanced browning similar to phosphate. For example a xylose/β-alanine reaction in succinate buffer (pK_2 5.6) at an initial pH of 7.14 developed an A_{420} of 0.249 in 80 min with a pH change of − 0.78. For comparison a similar reaction run in bis/tris buffer led to A_{420} of only 0.050. Under identical conditions using 0.1 M citrate buffer at an initial pH of 7.33 the xylose/β-alanine reaction developed an A_{420} of 0.379 (Δ pH = − 0.87). The effect of citrate on browning in a glucose-glycine model system has been observed to diminish relative to phosphate under milder reaction conditions (pH 7 / 25 °C) (5).

Whereas this study is concerned mainly with Maillard browning and non-volatile products formed near neutral pH, a recent report by Mottram and Norbrega described the accelerating effects of phosphate and phthalate ions on the formation of flavor volatiles in a cysteine/ribose system at acidic pH (8). The catalytic activity of phosphate at acid pH suggested the involvement of dihydrogen phosphate ion in the Maillard reaction mechanism.

Sugar Degradation by Buffer Anions

The effect of phosphate ion alone on sugars alone is illustrated in Figure 3. The three sugars that produced the greatest Maillard browning (ribose > xylose > galactose) browned to a lesser degree but with the same order of reactivity in the absence of β-alanine. Also, as found in the Maillard reactions, browning was virtually eliminated in the total absence of phosphate ion.

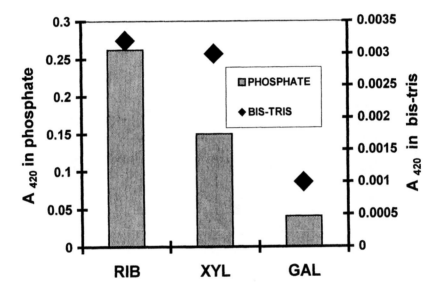

Figure 3. Browning of sugars in absence of amino acid after 100°C/80 min. in phosphate and bis-tris buffer at pH 7.3

The degree of ribose browning decreased markedly with lower phosphate concentration or in the case of a phosphate monoester. Under Figure 2 conditions reacting 0.1 M ribose for 80 min, 0.05 M phosphate in 0.10 M bis-tris buffer (initial pH 7.3) developed an A_{420} of only 0.045 compared to 0.262 obtained in 0.10 M phosphate. And, in a similar reaction, a phosphate monoester, DL-α-glycerophosphate at 0.05 M in 0.10 M bis-tris buffer (initial pH 7.3) gave a lower but comparable browning effect: A_{420} 0.026. A control experiment in 0.1 M bis-tris buffer produced no visible browning under the same conditions.

To gain insight into the effects of buffer anions on sugars alone we investigated reaction products for the presence of 1,2-dicarbonyl fragments which are known to be key intermediates for color formation in the Maillard reaction. The ribose/phosphate reaction was chosen as a model for study and o-phenylenediamine (OPD) was used to trap 1,2-dicarbonyls as their more stable and easily analyzed quinoxaline derivatives. The ribose/phosphate reaction mixture described in Figure 3 was treated with OPD and analyzed qualitatively by GC/MS. Four volatile quinoxalines were identified, namely: quinoxaline; 2-methyl; 2,3-dimethyl; and 2-ethylquinoxaline corresponding to dicarbonyls: ethanedial (glyoxal) **1**; 2-oxopropanal (pyruvaldehyde) **2**; 2,3-butanedione (diacetyl) **3** and 2-oxobutanal **4**. Based on quinoxaline peak areas the relative proportions of **1**, **2**, **3** and **4** were : 0.29, 1.00, 0.035 and 0.063 respectively. Also, a control reaction using bis/tris buffer instead of phosphate gave no detectable quinoxalines after OPD treatment. Clearly, phosphate ion accelerated the fragmentation of ribose into 1,2-dicarbonyls at neutral pH where the effect of hydrogen and/or hydroxide ion appeared to be negligible.

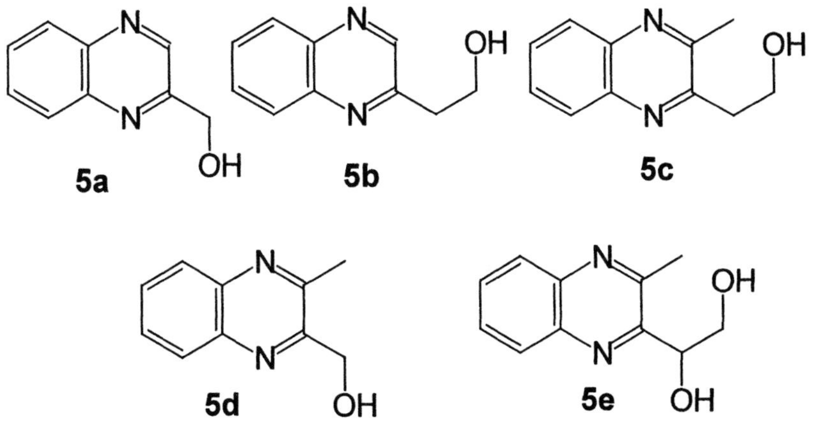

Figure 4. Quinoxaline derivatives of 1,2-dicarbonyls derived from ribose

The mode of ribose fragmentation was further investigated by TLC analysis of the less volatile quinoxaline products from the ribose/phosphate reaction.

Silica gel TLC indicated a major component in the R_f region previously known to contain quinoxaline derivatives of 5-carbon deoxyosones (9). A control experiment without phosphate produced none of this material. Preparative TLC in this R_f region provided a substance whose UV spectrum was consistent for a quinoxaline structure or mixture of quinoxalines (10). Analysis of the substance by 300 MHz proton NMR provided tentative evidence for the quinoxaline mixture **5a – 5e**, (Figure 4) based on comparisons with published NMR data (9-11). From peak area measurements the relative amounts present were: **5e** > **5c** ~ **5d** >> **5a** ~ **5b**. Approximately 70% of the mixture consisted of 2-methylquinoxaline derivatives based on peak area ratios. Quinoxalines **5a – 5e** are derived from a corresponding group of 1,2-dicarbonyl compounds, **6a – 6e** arising from ribose degradation, namely: 3-hydroxy-2-oxopropanal, **6a**; 4-hydroxy-2-oxobutanal, **6b**; 5-hydroxy-2,3-pentanedione, **6c**; 4-hydroxy-2,3-butanedione, **6d** and 4,5-dihydroxy-2,3-pentanedione, **6e**. Minor peaks in the quinoxaline mixture spectrum suggested that a trace of 2-oxo-4,5-dihydroxypentanal (3-deoxyribosone) **6f** may also have been present in the original reaction mixture.

A scheme depicting ribose degradation in terms of the <u>major</u> 1,2-dicarbonyl products observed is shown in Figure 5. Key to the formation of all products is the equilibration (isomerization) of aldose ↔ ketose sugars, in the case of ribose it is ribose ↔ ribulose. Sugar dehydration leads to the formation of deoxyosones, **6e** and **6f**. Retroaldol reaction (RA) of 3-deoxyosone **6f** explains the formation of pyruvaldehyde, **2**. The 1-deoxyosone **6e** may be the progenitor of additional dicarbonyls **3**, **4** and **6d**. RA of **6e** with loss of formaldehyde could generate **6d** directly whereas **3** and **4** require more circuitous routes. Isomerization of **6e** followed by dehydration can afford 3,4-dioxopentanal **7**, which by hydrolysis and loss of formic acid could produce **3**. The drop in pH during the ribose/phosphate reaction is consistent with formic acid formation and provides additional indirect evidence for **7**.

Compounds **1**, **4**, **6a**, **6b** and **6c** are chemically unique in that oxidation/reduction steps appear to be needed for their formation. Conceivably , hydroxyacetaldehyde from RA of ribose is oxidized (- 2H) to yield **1**. The unusual dicarbonyl **4** is probably formed by the aldol condensation/dehydration of hydroxyacetaldehyde and acetaldehyde. And, the acetaldehyde required for this reaction could derive from **7** following C-3 carbonyl reduction (+ 2H) and RA. A C-1 carbonyl reduction of **7** might also explain the origin of **6c**. Minor product **6a** may be derived via oxidation of **2** (+O) or from **6f** after oxidation at C-3 (+O) followed by RA reaction. Compound **6b** is conceivably derived from the 3-deoxyosone **6f**. Oxidation at C-3 of **6f** (+ O) [i.e., formation of ribosone] followed by fragmentation with loss of water and HCHO would afford a four carbon tricarbonyl (2,4-dioxobutanal) which upon reduction at C-4 could yield **6b**. Redox processes appear to be an integral part of Maillard chemistry,

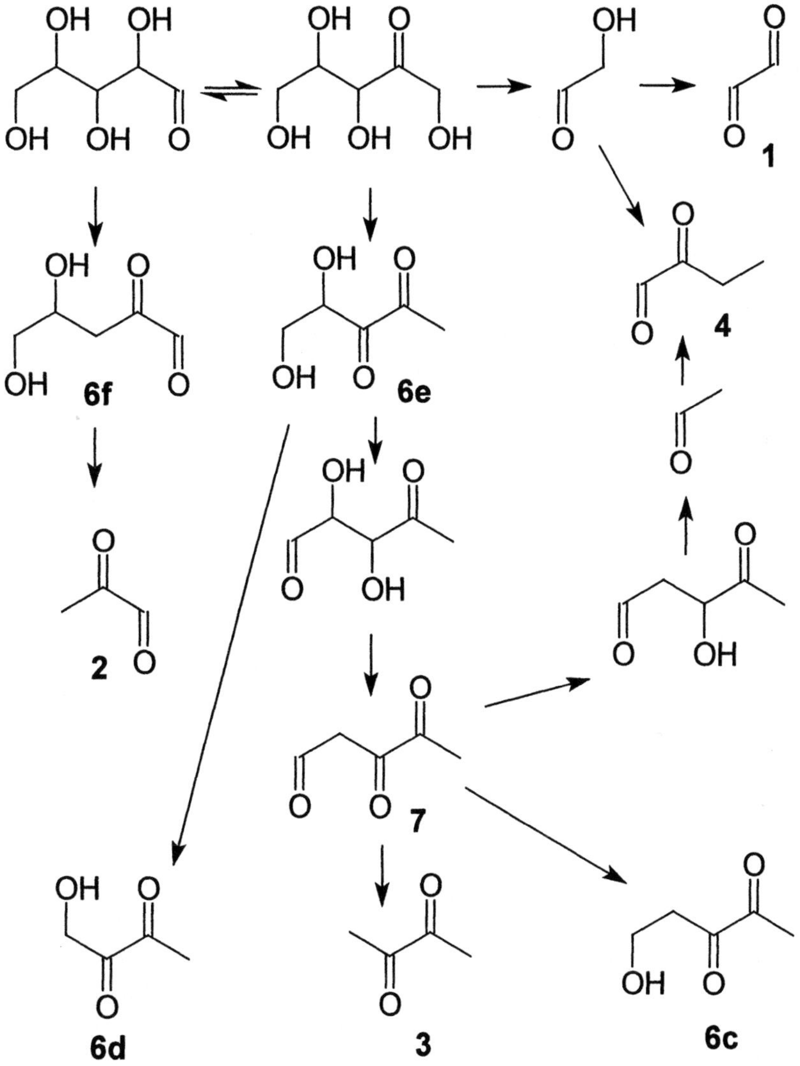

Figure 6. Reaction scheme for ribose degradation catalyzed by phosphate ion

however their mode of action remains a mystery and is therefore a subject worthy of future research.

Surprisingly, no definitive evidence was found for the relevant 3-deoxyosone **6f** except for its possible degradation products **2, 6a** and **6b**.

Mechanism of Catalytic Activity

The apparent enhanced basicity of polyatomic anions compared with hydroxyl ion is not obvious because electron delocalization in the polyatomic anions should make them very weak bases and therefore much less effective for proton withdrawal than hydroxyl ions.

Apart from a possible concentration effect due to high buffer strength compared to hydroxyl ion concentration near neutral pH the catalytic activity of carboxylate and dihydrogenphosphate ions can be explained by an intramolecular effect involving their common structural element: $O=X-O^- \leftrightarrow ^-O-X=O$ (abbreviated XO_2^-) where $X = RC$ for carboxylate and $X = P(OH)_2$ for dihydrogenphosphate (Figure 6). We suggest that catalytic activity begins with nucleophilic addition of XO_2- to the sugar carbonyl. For phosphate ions nucleophilic addition is well documented in catalysis of hydrolysis reactions (12). Addition of XO_2^- to a reducing sugar carbonyl group orients the $X = O$ function in a favorable position to abstract a proton from the α-carbon atom of the sugar leading either to enolization (path A) or dehydration (path B). Path A facilitates aldose $\leftrightarrow$ ketose isomerization whereas path B results in the formation of a 3-deoxyosone, e.g. **6f** in the case of ribose. An intrinsic catalytic effect of XO_2^- can therefore be attributed to the possibility of a more efficient (intramolecular) proton abstraction process compared to the intermolecular acid-base reaction otherwise available for anions including hydroxide ion. The same mechanism should apply for a wide variety of carboxylates, substituted phosphates and related arsenates. Actual intermediates involved in proton transfer is a matter for further speculation. At pH 7 phosphoric acid and carboxylic acids can probably be excluded because of their low pK values. A possible intermediate might be a zwitterionic species, i.e. $^-O-X^+-OH$ which later transfers a proton to enolate or hydroxide ions to regenerate the catalyst. In theory monohydrogen phosphate ion, HPO_4^{-2} might also act as a catalyst by following a similar mechanism. Our data does not exclude this possibility, however we favor the XO_2^- hypothesis because it uniquely includes the catalytic behavior of carboxylate ions.

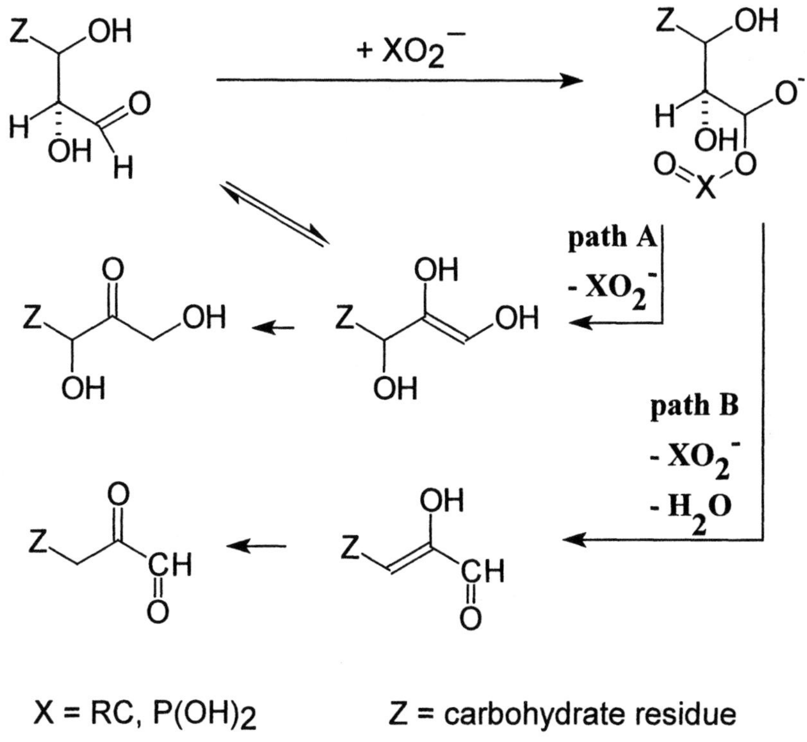

X = RC, P(OH)$_2$ Z = carbohydrate residue

Figure 6. Mechanism for catalysis of sugar degradation

In conclusion our experiments suggest that polyatomic anions of the form XO_2^- probably facilitate the aldose/ketose equilibration of reducing sugars and aldose sugar dehydration to form 3-deoxyosones. Subsequent reactions apparently lead to additional α-dicarbonyl formation. And, because α-dicarbonyls are key intermediates that lead to colored melanoidins the extra α-dicarbonyls arising from XO_2^- catalyzed sugar degradation explains the enhanced browning observed overall in XO_2^- catalyzed Maillard reactions.

168

Acknowledgements

This paper is dedicated to the memory of my late friend and colleague Dr. Paul Borkowski whose work on microwave-induced browning of foods provided the inspiration for this study. Also, I thank the Procter & Gamble Co. for allowing me to pursue independent research in their laboratories following my retirement. And, I thank Dr. Gerhard Zehentbauer and Mrs. Molly Armstrong for performing NMR measurements and Mr. Don Patton for technical assistance with GC/MS.

References

1. Martins, S. I. F. S.; Jongen, W. M. F.; van Boekel, M., A., J., S. *Trends in Food Sci. and Technol.* **2001**, 11, 364373.
2. Reynolds, T. M. *Aust. J. Chem.* **1958**, 11, 265-274.
3. Potman, R. P.; van Wijk, Th. A. In *Thermal Generation of Aromas*; Parliament, T. H.; McGorrin, R. J.; Ho, C.-T., Eds. ACS Symposium Series 409; American Chemical Society: Washington, DC, 1989, pp182-195.
4. Davidek, T.; Clety, N.; Aubin, S.; Blank, I. *J. Agric. Food Chem.* **2002**, 50, 5472-5479.
5. Bell, L. N. *Food Chemistry* **1997**, 59, 143-147.
6. Weenen, H.; Apeldoorn, W. In *Flavour Science Recent Developments*; Taylor, A. J.; Mottram, D. S., Eds., RSC Publishing: Cambridge, U.K., **1996**, pp 211-216.
7. Angyal, S. J. *Adv. Carbohydr. Chem. Biochem.* **1984**, 42, 15-68.
8. Mottram, D.S.; Norbrega, I.C.C. *J. Agric. Food Chem.* **2002**, 50, 4080-4086.
9. Nedvidek, W.; Ledl, F.; Fischer, P. *Z. Lebensm. Unters. Forsch.* **1992**, 194, 222-228.
10. Morita, N.; Daido, Y.; Takagi, M. *Agric. Biol. Chem.* **1984**, 48, 3161-3163.
11. Hollnagel, A.; Kroh, L. W. *J. Agric. Food Chem.* **2002**, 50, 1659-1664.
12. Bender, M. L. In *Mechanisims of Homogeneous Catalysis from Protons to Proteins*, Wiley-Interscience, New York, NY, **1971**, p157.

Analysis of Process Flavors

Chapter 14

Synthesis and Sensorial Description of New Sulfur-Containing Odorants

F. Robert[1], H. Simian[1], J. Héritier[1,2], J. Quiquerez[1], and I. Blank[1]

[1]Nestec Ltd., Nestlé Research Center, Vers-chez-les-Blanc,
1000 Lausanne 26, Switzerland
[2]Current address: Haute Ecole Valaisanne, 1950 Sion 2, Switzerland

Various sulfur-containing odorants, such as 2-thioalcanes, 3-acetylthio-2-alkyl alkanals, and trialklylated 1,3,5-dithiazines were prepared by conventional, parallel, and split-mix synthesis approaches. 2-Heptanethiol, newly identified in bell peppers, and 3-acetylthio-2-methylpentanal showed relatively low odor thresholds of 10 and 5 µg/kg water, respectively. Several odorants were found to develop interesting notes, which are compatible with both savory and sweet flavors.

Introduction

Volatile organic sulfur compounds contribute to the aroma of many vegetables, fruits, and food products (*1, 2*). In general, thiols belong to the most intense and characteristic aroma substances with sulfury, vegetable-like, fruity notes perceived at low concentrations. In bell peppers, we have recently identified 2-heptanethiol as an odor-active compound (*3*). However, the identification of thiols is generally a challenging task due to their instability and the low concentration. The sensory relevance of such odorants can be explained by their low threshold values. For example, (2*R*,3*S*)-3-mercapto-2-methyl-1-pentanol, a character-impact constituent of fresh onions, shows an odor threshold of 0.03 µg/kg water (*4*).

Identification of sensory relevant compounds can be achieved by applying a combinatorial approach. Vermeulen and coworkers (*5-8*) have prepared a series of sulfur-containing odorants by reacting various precursors in one reaction vessel. Odor-active compounds were screened by gas the chromatography-sniffing techniques for identification experiments, which are facilitated due to higher concentrations as compared to natural extracts.

An alternative approach is to run several reactions in parallel in different reactors, a method called parallel synthesis (*9*). In general, combinatorial approaches are easy to run and could be applied for the identification of unknown compounds. However, the analysis is more complex since mixtures of many products are obtained. The parallel approach has the advantage of simple automation leading to several isolated compounds.

This article deals with the combinatorial synthesis of various sulfur-containing odorants. We used this approach for the identification of natural aroma components in different ingredients such as fried onion or bell peppers.

Experimental

Materials

The following chemicals were commercially available and of highest purity: piperidine, 2-hexanol, 2-heptanol, 2-octanol, 2-nonanol, ethanal, isobutanal, *p*-toluenesulfonyl chloride, sodium hydrogensulfide monohydrate ($NaSH \cdot H_2O$), ammonium sulfide, deuterochloroform (C^2HCl_3) (Fluka/Aldrich, Buchs, Switzerland); hydrochloric acid (HCl, 37%), magnesium sulfate ($MgSO_4$), silica gel 60 (Merck, Darmstadt, Germany); citric acid monohydrate (Citrique Belge, Belgium). The solvents pentane, dimethylformamide (DMF), ethyl acetate (EtOAc), toluene, and diethyl ether (Et_2O) were from Merck and freshly distilled prior to use. Solvents were distillated on sodium/benzophenone (Fluka/Aldrich) for air-sensitive reactions, which were carried out under nitrogen atmosphere. Purification by column chromatography was carried out on silica gel 60 (Fluka).

Synthesis of 2-Thioalcanes

The aliphatic 2-thioalcanes were synthesized from the corresponding 2-alcanols as starting material using classical approaches (*10*), i.e. by tosylation of the alcohol and nucleophilic substitution of the intermediary tosylate with sodium hydrogensulfide into the target thiol.

172

General Procedure for the Synthesis of 2-(p-Toluenesulfonyl)alcanes. These were obtained from the corresponding 2-alcanols **1a-d** (200 mmol, 1.0 equiv) dissolved in pyridine (135 mL) and cooled to 0 °C. *p*-Toluenesulfonyl chloride (41.8 g, 220 mmol, 1.1 equiv) was slowly added and the mixture was stirred at room temperature overnight. Toluene was added (200 mL), the reaction mixture filtered, and the filtrate washed with toluene (200 mL). The mother liquor was washed twice with an aqueous HCl solution (5 N, 200 mL). The organic layer was dried over MgSO$_4$ and concentrated under reduced pressure. After dry chromatography on silica gel (pentane/EtOAc, 9:1, v/v), the 2-(*p*-toluenesulfonyl)alcanes **2a-d** were obtained as colorless oils in 60-80% yields.

2-(*p*-Toluenesulfonyl)hexane **2a**. Yield: 73%. ^{1}H NMR (360 MHz, C^{2}HCl$_3$, δ/ppm): 0.74 (t, 3H, CH$_3$, ^3J = 7.1 Hz), 1.05-1.34 (m, 4H, 2 CH$_2$), 1.19 (d, 3H, CH$_3$, ^3J = 6.1 Hz), 1.38-1.52 (m, 2H, CH$_2$), 2.37 (s, 3H, CH$_3$), 4.58 (qt, 1H, CH, ^3J = 6.1 Hz, ^3J = 6.35 Hz), 7.28 (d, 2H, ^3J = 8.3 Hz), 7.73 (d, 2H, ^3J =8.3 Hz); ^{13}C NMR (90 MHz, C^{2}HCl$_3$, δ/ppm): 14.2 (CH$_3$), 21.2 (CH$_3$), 22.0 (CH$_3$), 21.9 (CH$_3$), 22.6 (CH$_2$), 27.3 (CH$_2$), 36.5 (CH$_2$), 81.0 (O-CH), 128.0 (C=CH), 130.1 (C=CH), 134.9 (C=C), 144.8 (C=C).

2-(*p*-Toluenesulfonyl)heptane **2b**. Yield: 76%. ^{1}H NMR (360 MHz, C^{2}HCl$_3$, δ/ppm): 0.82 (t, 3H, CH$_3$, ^3J = 7.2 Hz), 1.12-1.23 (m, 6H, 3 CH$_2$), 1.26 (d, 3H, CH$_3$, ^3J = 6.3 Hz), 1.42-1.63 (m, 2H, CH$_2$), 2.44 (s, 3H, CH$_3$), 4.59 (qt, 1H, CH, ^3J = 6.1 Hz, ^3J = 6.35 Hz), 7.33 (d, 2H, ^3J = 8.3 Hz), 7.79 (d, 2H, ^3J = 8.3 Hz). ^{13}C NMR (90 MHz, C^{2}HCl$_3$, δ/ppm): 14.3 (CH$_3$), 21.3 (CH$_3$), 22.1 (CH$_3$), 22.8 (CH$_2$), 24.9 (CH$_2$), 31.7 (CH$_2$), 36.8 (CH$_2$), 81.1 (O-CH), 128.1 (C=CH), 130.1 (C=CH), 135.0 (C=C), 144.8 (C=C).

2-(*p*-Toluenesulfonyl)octane **2c**. Yield: 75%. ^{1}H NMR (360 MHz, C^{2}HCl$_3$, δ/ppm): 0.90 (t, 3H, CH$_3$, ^3J = 7.0 Hz), 1.20-1.36 (m, 8H, 4 CH$_2$), 1.31 (d, 3H, CH$_3$, ^3J = 6.35 Hz), 1.47-1.70 (m, 2H, CH$_2$), 2.49 (s, 3H, CH$_3$), 4.65 (qt, 1H, CH, ^3J = 6.1 Hz, ^3J = 6.35 Hz), 7.38 (d, 2H, ^3J = 8.1 Hz), 7.85 (d, 2H, ^3J = 8.1 Hz). ^{13}C NMR (90 MHz, C^{2}HCl$_3$, δ/ppm): 14.4 (CH$_3$), 21.3 (CH$_3$), 22.0 (CH$_3$), 22.9 (CH$_2$), 25.2 (CH$_2$), 29.2 (CH$_2$), 31.9 (CH$_2$), 36.9 (CH$_2$), 81.1 (O-CH); 128.1 (C=CH), 130.1 (C=CH), 135.0 (C=C), 144.8 (C=C).

2-(*p*-Toluenesulfonyl)nonane **2d**. Yield: 62%. ^{1}H NMR (360 MHz, C^{2}HCl$_3$, δ/ppm): 0.83 (t, 3H, CH$_3$, ^3J = 6.8 Hz), 1.05-1.34 (m, 10H, 5 CH$_2$), 1.27 (d, 3H, CH$_3$, ^3J = 6.2 Hz), 1.35-1.60 (m, 2H, CH$_2$), 2.39 (s, 3H, CH$_3$), 4.54 (qt, 1H, CH, ^3J = 6.2 Hz, ^3J = 6.35 Hz), 7.28 (d, 2H, ^3J = 8.3 Hz), 7.74 (d, 2H, ^3J = 8.3 Hz). ^{13}C NMR (90 MHz, C^{2}HCl$_3$, δ/ppm): 14.5 (CH$_3$), 21.3 (CH$_3$), 22.0 (CH$_3$), 23.0 (CH$_2$), 25.3 (CH$_2$), 29.4 (CH$_2$), 29.5 (CH$_2$), 32.1 (CH$_2$), 36.9 (CH$_2$), 81.1 (O-CH), 128.1 (C=CH), 130.1 (C=CH), 135.0 (C=C), 144.8 (C=C).

General Procedure for the Synthesis of 2-Thioalcanes. 2-(*p*-Toluenesulfonyl)alcanes **2a-d** (37 mmol, 1.0 equiv) and NaSH·H$_2$O (7.0 g, 94 mmol, 2.5 equiv) were stirred in DMF (25 mL) at 80 °C for 2 h. The reaction mixture was diluted in brine (200 mL) and the aqueous layer was extracted with

Et$_2$O (3 x 200 mL). The organic layers were combined and washed with brine (5 x 200 mL), dried over MgSO$_4$ and concentrated under reduced pressure. The racemic target compounds **3a-d** was obtained after distillation under reduced pressure in 20-40% yield.

2-Hexanethiol **3a**. ^{1}H NMR (360 MHz, C^{2}HCl$_3$, δ/ppm): 0.89 (t, 3H, CH$_3$, ^3J = 7.2 Hz), 1.27-1.36 (m, 6H, 3 CH$_2$), 1.36 (d, 3H, CH$_3$, ^3J = 6.7 Hz), 1.49-1.62 (m, 2H, CH$_2$), 2.96 (tq, 1H, ^3J = 6.5 Hz, ^3J = 6.1 Hz). ^{13}C NMR (90 MHz, C^{2}HCl$_3$, δ/ppm): 14.4 (CH$_3$), 22.8 (CH$_2$), 26.0 (CH$_3$), 30.0 (CH$_2$), 36.0 (CH$_3$), 41.1 (CH). MS (EI, m/z, rel-%): 118 (M$^+$, 35), 85 (26), 84 (26), 69 (33), 61 (65), 56 (42), 55 (40), 41 (100). GC: RI(PONA) = 850, RI(DB-Wax) = 1063. Yield: 20%, boiling point: 34°C (14 mbar).

2-Heptanethiol **3b**. ^{1}H NMR (360 MHz, C^{2}HCl$_3$, δ/ppm): 0.82 (t, 3H, CH$_3$, ^3J = 7.2 Hz), 1.27-1.38 (m, 6H, 3 CH$_2$), 1.36 (d, 3H, CH$_3$, ^3J = 6.7 Hz), 1.49-1.62 (m, 2H, CH$_2$), 2.96 (tq, 1H, ^3J = 6.5 Hz, ^3J = 6.1 Hz). ^{13}C NMR (90 MHz, C^{2}HCl$_3$, δ/ppm): 14.5 (CH$_3$), 23.0 (CH$_2$), 26.0 (CH$_3$), 27.5 (CH$_2$), 31.9 (CH$_2$), 36.0 (CH$_3$), 41.3 (CH). MS (EI, m/z, rel-%): 132 (M$^+$, 32), 98 (31), 70 (32), 69 (26), 61 (60), 57 (100), 56 (74), 55 (40), 43 (41), 41 (76). GC: RI(PONA) = 953, RI(DB-Wax) = 1170. Yield: 40%, boiling point: 120°C (143 mbar).

2-Octanethiol **3c**. ^{1}H NMR (360 MHz, C^{2}HCl$_3$, δ/ppm): 0.90 (t, 3H, CH$_3$, ^3J = 7.2 Hz), 1.29-1.60 (m, 13H, 1 CH$_3$, 5 CH$_2$), 2.94 (tq, 1H, ^3J = 7.0 Hz, ^3J = 6.1 Hz). ^{13}C NMR (90 MHz, C^{2}HCl$_3$, δ/ppm): 14.5 (CH$_3$), 23.0 (CH$_2$), 26.0 (CH$_3$), 27.8 (CH$_2$), 29.4 (CH$_2$), 32.2 (CH$_2$), 36.0 (CH$_3$), 41.3 (CH$_2$). MS (EI, m/z, rel-%): 146 (M$^+$, 16), 112 (27), 84 (22), 83 (34), 71 (45), 70 (58), 61 (64), 57 (64), 56 (48), 55 (68), 43 (62), 41 (100). GC: RI(PONA) = 1057, RI(DB-Wax) = 1271. Yield: 24%, boiling point: 44°C (13 mbar).

2-Nonanethiol **3d**. ^{1}H NMR (360 MHz, C^{2}HCl$_3$, δ/ppm): 0.82 (t, 3H, CH$_3$, ^3J = 7.2 Hz), 1.27-1.38 (m, 15H, 6 CH$_2$), 2.96 (tq, 1H, ^3J = 7.0 Hz, ^3J = 6.1 Hz). ^{13}C NMR (90 MHz, C^{2}HCl$_3$, δ/ppm): 14.5 (CH$_3$), 23.0 (CH$_2$), 26.0 (CH$_3$), 27.9 (CH$_2$), 29.6 (CH$_2$), 29.7 (CH$_2$), 32.2 (CH$_2$), 36.0 (CH$_3$), 41.3 (CH). MS (EI, m/z, rel-%): 160 (M$^+$, 17), 126 (31), 97 (31), 85 (22), 84 (28), 70 (40), 69 (42), 61 (58), 56 (62), 55 (76), 43 (77), 41 (100). GC: RI(PONA) = 1160, RI(DB-Wax) = 1377. Yield: 34%, boiling point: 28°C (0.03 mbar).

Parallel Synthesis of 3-Acetylthio-2-alkyl Alkanals

This was achieved using the Quest 205 apparatus from Argonaut Technologies (Basel, Switzerland). Combination of parallel synthesis with on-line work-up and sample collection allows synthesis procedures using one single instrument. Each reaction vessel has a port at the top, a drain valve at the bottom and inert gas to control draining time. This gives a number of options for on-line work-up such as liquid-liquid extraction or solvent evaporation.

Synthesis of 3-Acetylthio-2-alkyl Alkanals. Piperidine (100 µL) was added to alkenals **5a-f** (34 mmol) under nitrogen at 10 °C in separated cylinders of the Quest 205 apparatus. Thioacetic acid (3.68 mL, 51.6 mmol) was then added drop-wise at 10 °C. Thereafter, the reaction mixture was stirred for another 18 h at room temperature. The mixture was diluted with Et_2O (10 mL), washed first with HCl (10 mL, 1 N) and then twice with a saturated $NaHCO_3$ solution (10 mL). The organic phases were dried over Na_2SO_4. All these steps were carried out at the same time in the Quest 205. Then, the solvent was evaporated for each sample. The GC purity of the crude products was 50-90%, depending on the starting alkenal. In each case, a mixture of the two diastereomers was obtained. The detailed analytical description of the 3-acetylthio-2-alky-alkanals **6a-f** has been reported elsewhere (*11*).

Split-mix Synthesis of Trialkylated 1,3,5-Dithiazines

In a glass reactor containing an aqueous acid solution of pH 3-4, isobutanal (7.10 mL) was added slowly followed by ethanal (8.65 mL). Then, an ammonium sulfide solution (44.6 mL, 21%) was added dropwise, keeping the temperature in the reactor at about 2-3 °C. After the addition, the reaction mixture was stirred for 30 min at 2-3°C before warming up to 20°C. The sample was stirred for additional 12 h at 20°C. The aqueous phase was removed and the organic phase filtered to obtain a yellow oil (11.8 g), which was stirred under vacuum (30 mbar) at 60°C for 3 h. Distillation of the clear yellow oil (10.7 g) under reduced pressure (1 mbar) at 70-120°C led to a yellow oil (8.74 g, 60% yield) with strong odor.

Gas Chromatography - Mass Spectrometry / Olfactometry (GC-MS/O)

Mass spectra of the synthesized compounds and their retention indices were acquired using a gas chromatograph GC 5890 (Agilent, Geneva, Switzerland) equipped with 2 splitless injectors heated at 260 °C and coupled with a quadrupole mass spectrometer MS 5970 (Agilent, Geneva, Switzerland) operating in the electron impact ionization mode at 70 eV. Acquisitions were carried out over a mass range of 10 to 350 Da. Separations were performed on a 100% dimethyl polysiloxane apolar stationary phase (Ultra-1 PONA, 50 m x 0.20 mm i.d., 0.5 µm film thickness, Agilent) and on a polyethylene glycol polar stationary phase (DB-Wax, 60 m x 0.25 mm i.d., 0.5 µm film thickness, J&W, Folsom, CA). Helium was used as the carrier gas with a constant flow rate of 0.6 mL/min and 1.0 mL/min, respectively. The oven was programmed as follows: 20 °C (0.5 min), 70 °C/min to 60 °C, 4 °C/min to 240 °C. The temperature of the transfer line was held at 280 °C during the chromatographic run. Sniffing detection was performed on both stationary phases.

Nuclear Magnetic Resonance (NMR) Spectroscopy

The samples for NMR spectroscopy were prepared in Wilmad 528-PP 5 mm Pyrex NMR tubes using deuterochloroform as solvent (0.7 mL). The NMR spectra (^{1}H-NMR, ^{13}C-NMR, DEPT-135) were acquired on a Bruker AM-360 spectrometer equipped with a quadrinuclear 5 mm probe head, at 360.13 MHz for ^{1}H and at 90.03 MHz for ^{13}C under standard conditions (*12*). All chemical shifts are cited in ppm relative to the solvent signal.

Threshold Determination

Orthonasal detection thresholds were determined in water (Vittel) with seven panelists. Eight samples were presented in order of decreasing concentrations (factor 10 between samples) to estimate the range of threshold concentration. Threshold values were determined by triangular test using a series of four concentrations (factor 2.5 between samples). Threshold values correspond to ≥70% of correct answers.

Results and Discussion

2-Thioalcanes

Synthesis. In the frame of our work on the identification of impact odorants found in bell peppers and onions (*3*), we synthesized various secondary thiols using the pathway described in Figure 1. The alcohols **1a-d** were activated with *p*-toluenesufonyl chloride as a good leaving group. This reaction was performed with 2-hexanol **1a**, 2-heptanol **1b**, 2-octanol **1c**, and 2-nonanol **1d**. The different tosyl derivatives were obtained in good 73%, 76%, 75% and 62% yield, respectively. The tosylate group was then substituted with sodium hydrogen sulfide as nucleophile. The conversion yields were excellent, however the isolated yields were moderate (20-40%), mainly due to the instability of the thiols, which were easily oxidized to the corresponding disulfide during the distillation procedure, as indicated by GC analysis (data not shown).

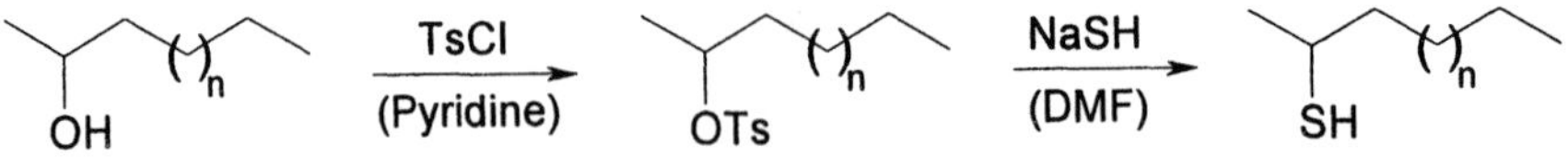

Figure 1. Synthesis pathway to 2-thioalcanes

Sensory Properties. The odor properties of the 2-thioalcanes synthesized are summarized in Table 1. The odor thresholds determined in water were between 10 to 90 μg/L. 2-Heptanethiol showed the lowest orthonasal detection threshold value (10 μg/L in water). This threshold is similar to hydrogen sulfide having a sulfury, egg-like note (*2*). In high dilution, *i.e.* about 10-times the threshold, 2-heptanethiol was mainly described as bell pepper-like, fruity, and vegetable-like. At higher concentrations (100-1000 times the threshold), the odor description was completely different: sulfury, onion-like, with some mushroom note. In the literature, the odor of 2-heptanethiol has been described as sweet, fruity, tropical (sulfury), and floral, however without indicating the threshold value (*13*).

Table I. Sensorial Properties of 2-Thioalcanes.

Compound	Odor description[a]	Odor threshold[b]
2-Hexanethiol **3a**	Sulfury, leek	60
2-Heptanethiol **3b**	Bell pepper, green vegetables, sulfury[a], onion[a], mushroom[a]	10
2-Octanethiol **3c**	Mushroom, slightly fruity, bell pepper	50
2-Nonanethiol **3d**	Herbaceous, mushroom, woody	90

[a] Odor description at a higher concentration (about 1 mg/L). [b] Detection threshold in water (μg/L) was obtained by orthonasal measurements performed by 7 panelists.

2-Hexanethiol **3a** was described as more leek-like. On the contrary, the higher molecular weight odorants, 2-octanethiol **3c** and 2-nonanethiol **3d**, were reminiscent of mushroom and herbaceous sensory characteristics, which is close to sensory properties of the corresponding C8 and C9 alcohols known to elicit mushroom notes (*14*). The descriptors of the 2-thioalcanes are in accordance with those reported by Sakoda and coworkers (*13*) who studied the relationship of odor and chemical structure comparing 1- and 2-alkyl alcohols and their corresponding thiols. To our knowledge, none of the thioalcanes studied have been found in natural products or reported as constituents of food.

3-Acetylthio-2-alkyl Alkanals

Parallel synthesis. Several 3-acetylthio-2-alkyl alkanals with different chain length were prepared by parallel synthesis. As shown in **Figure 2**, the first step was an aldol-type condensation of an n-alkanal, which results in the corresponding 2-alkyl-α,β-unsaturated aldehyde. Depending on the starting aldehyde, the yields varied between 40 and 70%. The alkenals were then mixed with thioacetic acid under alkaline condition to introduce regioselectively the acetylthio function in the aldehyde backbone. Structure characterization and purity control was carried out on the basis of GC, GC-MS, and NMR data. The purity (GC) of the crude products was between 50 and 90% and conversion yields were between 55 and 98%, depending on the starting alkenal. The analytical data are reported elsewhere (*11*).

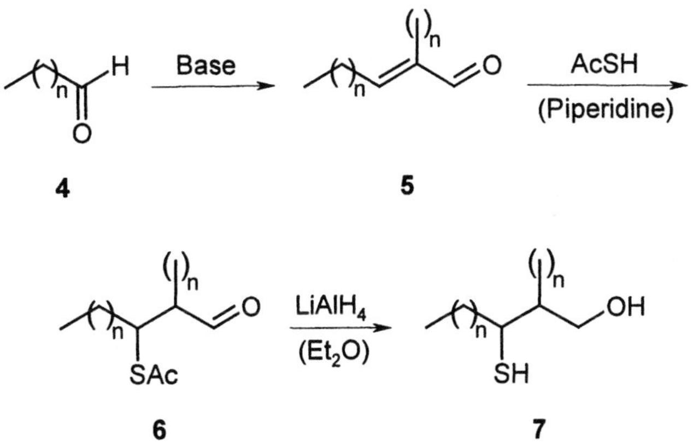

Figure 2. Synthesis pathway to 2-alkyl-3-mercapto alcohols

To obtain the corresponding mercaptoalcohols, both the acetylthio and aldehyde functions can be reduced with lithium aluminum hydride under inert conditions (*7*, Figure 1). Also the mercaptoalcohols were obtained as a mixture of diastereomers. In general, conversion yields were about 80 % in all cases as indicated by GC. Only 3-mercapto-2-methyl-1-pentanol was isolated and characterized by GC-MS and NMR. The data were in good agreement with those reported in the literature (4).

Sensory Properties. The sensory properties of the acetylthio aldehydes are shown in **Table 2.** They are predominantly characterized by fruity notes, such as tropical and grapefruit-like, in particular odorants **6b-e.** On the contrary, the lower and higher homologues, i.e. 3-acetylthio-2-methyl pentanal **6a** and 3-acetylthio-2-hexyl decanal **6f**, showed savory characters such as onion-like and fatty, chicken-like, respectively. Surprisingly, 3-acetylthio-2-butyl octanal **6d** was described as fresh, green, vegetable-like, reminiscent of crude cauliflower, when it is smelt neat. From a structural point of view, these data are in good agreement with the tropical/vegetable "odorophore" recently proposed by Rowe and Tangel (*15*). Following our results, the "olfactophore" model based on a 1,3-oxygen-sulfur structure can further be generalized by adding the acetyl group for substituent A in Figure 3 (*16*). Interestingly, the threshold concentration of the 3-acetylthio-2-alkyl alkanals **6a-e** increased exponentially with the carbon chain length (*11*).

Table 2. Sensorial Properties of 3-Acetylthio-2-alkyl Alkanals

Compound	Odor description[a]	Odor threshold[b]
3-Acetylthio-2-methyl pentanal 6a	Leek, onion, bouillon	5
3-Acetylthio-2-ethyl hexanal 6b	Fruity, tropical, grapefruit	15
3-Acetylthio-2-propyl heptanal 6c	Fruity, tropical, grapefruit	50
3-Acetylthio-2-butyl octanal 6d	Fruity, grapefruit, green vegetables, cauliflower	200
3-Acetylthio-2-pentyl nonanal 6e	green, fruity, citrus	500
3-Acetylthio-2-hexyl decanal 6f	Fatty, chicken	n.d.

[a] Odor description was performed at a concentration of about 0.5-5 mg/L. [b] Detection threshold in µg/L water was obtained by orthonasal measurements performed by 7 panelists (n.d., not determined).

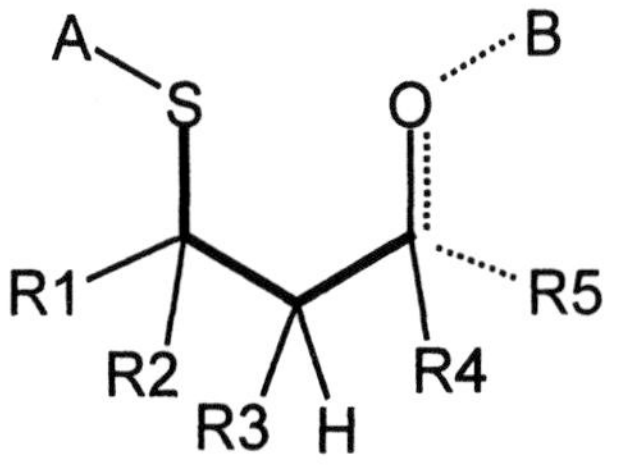

Figure 3. The "olfactophore" for tropical/vegetable notes with A: H, SCH₃, ring; B: H, CH₃, acyl, absent if carbonyl; R1, R2: H, alkyl; R3: H, alkyl, ring; R4: H, CH₃, ring, OR; R5: H, absent if carbonyl. Adapted from ref. 16.

Dihydro-2,4,6-trialkyl-4*H*-1,3,5-dithiazines

Synthesis (split-mix). Trialkylated dihydro-4*H*-1,3,5-dithiazines **10** were obtained by mixing aldehydes **8** and ammonium sulfide **9** in aqueous solution under acidic conditions (Figure 4). When using several aledhydes a mixture of dithiazines was obtained representing the different combinations between the reagents. For example, when using isobutanal and ethanal as starting aldehydes the six following 1,3,-5-dithiazine derivatives can theoretically be obtained: 2,4,6-trimethyl, 2-isopropyl-4,6-dimethyl, 4-isopropyl-2,6-dimethyl, 2,4-diisopropyl-6-methyl, 4,6-diisopropyl-2-methyl, 2,4,6-triisopropyl with a statistical distribution of 1/1/2/2/1/1. In practice, steric hindrance and electronic differences modified the statistical distribution leading to a 4/2/3/0.2/0.4/0.1 ratio of the above-mentioned dithiazine derivatives. In more general terms, n aldehydes should theoretically provide $n^2(n+1)/2$ dithiazines in the same reaction.

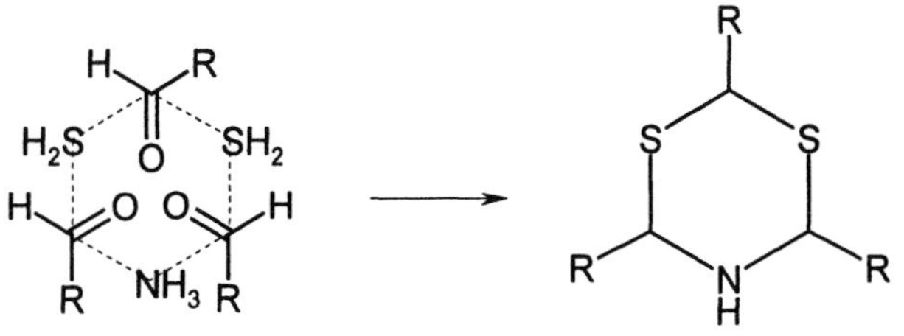

8, 9   **10**

Figure 4. Synthesis pathway to trialkylated 1,3,5-dithiazines

180

Sensory Properties. This mixture showed a very strong onion and leek-like aroma. All components can be sensorially characterized by GC-O. The sensory characteristics of the 1,3,5-dithiazines from isobutanal and ethanal used as starting aldehydes have already been reported earlier (*17*).

In conclusion, parallel synthesis and more generally, combinatorial chemistry, seems to be an attractive approach in aroma research to help identifying new odorants with interesting sensory properties. It allows rapid synthesis of a large number of components in a reasonable time. The reference compounds can be very useful for structure elucidation of unknown odorants.

References

1. Boelens, M. H.; van Gemert, L. J. *Perfumer & Flavorist* **1993**, *18(3)*, 29-39.
2. Blank, I. In *Heteroatomic Aroma Compounds*, ACS Symposium Series 826; Reineccius G.A., Reineccius T.A., Eds.; American Chemical Society: Washington, DC, 2002; pp 25-53.
3. Simian, H.; Robert, F.; Blank, I. *J. Agric. Food Chem.* **2004**, *52*, 306-310.
4. Widder, S.; Sabater Lüntzel, C.; Dittner, T.; Pickenhagen, W. *J. Agric. Food Chem.* **2000**, *48*, 418-423.
5. Vermeulen, C.; Pellaud, J. ; Gijs, L. ; Collin, S. *J. Agric. Food Chem.* **2001**, *49*, 5445-5449.
6. Vermeulen, C.; Collin, S. *J. Agric. Food Chem.* **2002**, *50*, 5654-5659.
7. Vermeulen, C.; Collin, S. *J. Agric. Food Chem.* **2003**, *51*, 3618-3622.
8. Vermeulen, C.; Guyot-Declerck, C.; Collin, S. *J. Agric. Food Chem.* **2003**, *51*, 3623-3628.
9. Gillmor, S. A.; Cohen, F. E. *Receptor* **1993**, *3(3)*, 155-163.
10. Fieser L. F.; Fieser M. *Reagents for Organic Synthesis*, Vol. 1; Wiley: New York, 1967.
11. Robert, F.; Héritier, J.; Quiquerez, J.; Simian, H.; Blank, I. *J. Agric. Food Chem.* **2004**, *52*, Submitted.
12. Lin, J.; Welti, H. D.; Arce Vera, F.; Fay, L. B.; Blank, I. *J. Agric. Food Chem.* **1999**, *47*, 2813-2821.
13. Sakoda, Y.; Hayashi, S. In *Advances in Flavours and Fragrances – From the Sensation to the Synthesis*; Swift, K. A. D., Ed.; Special Publication no. 277; Royal Society of Chemistry: Cambridge, 2002; pp 15-24.
14. Dijkstra, F. Y.; Wilken, T. O. *Z. Lebensm. Unters. Forsch.* **1976**, *160*, 255-262.
15. Rowe, D. J.; Tangel, B. *Perfumer & Flavorist* **1999**, *24(6)*, 36-41.
16. Rowe, D. J. In *Advances in Flavours and Fragrances – From the Sensation to the Synthesis*; Swift K. A. D., Ed.; Special Publication no. 277; The Royal Society of Chemistry: Cambridge, 2002; pp 202-226.
17. Werkhoff, P. Güntert, M.; Hopp, R. *Food Rev. Intern.* **1992**, *8*, 391-442.

Chapter 15

On-Line Monitoring of the Maillard Reaction Using a Film Reactor Coupled to Ion Trap Mass Spectrometry

G. A. Channell and A. J. Taylor

Division of Food Sciences, University of Nottingham, Sutton Bonington Campus, Loughborough LE12 5RD, United Kingdom

A micro-reactor has been developed which utilizes a continuous stream of carrier gas with controlled humidity to pass over a thin film of glucose and leucine as representatives of Maillard reactants. The efflux from the reactor interfaces directly to the APcI source of an Ion Trap Mass Spectrometer to record the protonated molecular ion of the species generated. Controlled temperature change within the reaction film under a 'stepped' regime 60°C→100°C→130°C→170°C at 1°C /s demonstrated the rapid production of 1-amino propane-2-one with increasing humidity. Tentative assignment of other associated molecular ions show a range of sugar breakdown products, amino acid/sugar adducts, transamination products, and the Strecker aldehyde, 3-methylbutanal to be produced. Reaction cores, removed from the micro-reactor and characterized for absorbance in methanolic solution at 450nm show increased coloration with increased humidity. Amino acid and sugar type have a greater influence on color formation than process conditions.

The Maillard reaction originating from the nucleophilic addition of α-amino acid to reducing sugar, involves a complex series of interconnected reactions with multiple intermediates and end products (1). The reaction environment is central to determining which reactions predominate and dictate the end product profile influencing the color and aroma of thermally processed foods (2). Many food processes involve short time, high temperature, low moisture conditions typically employed in frying/baking and represent a particular challenge to the analytical chemist in the chemical characterization of the reaction. Building on previous work from our laboratory (3,4), the purpose of the present investigation is to devise an analytical system which will allow the rapid on-line monitoring of a multitude of Maillard volatiles in short time scales with the capability to control secondary reactant species during the time course of the reaction, for example water, ammonia and hydrogen sulfide. The latter species are typically available as by-products of amino acid degradation during thermal processing and add to the complexity of the Maillard reaction.

The premise is to utilise a liquid film to provide a reaction environment which can be dynamically controlled in terms of heat and mass flux (influx/efflux) and to complement this with the on-line monitoring technique of Atmospheric Pressure chemical Ionisation (APcI)-Ion Trap Mass Spectrometry (ITMS). This technique allows the flux of protonated molecular ions (MH^+) to be directly monitored (mass spectral dimension 1) and to fragment these species under tailored conditions within the ion trap (Collision Induced Dissociation (CID),mass spectral dimension 2), to produce fragment ions representative of the parent ion. This capability is central to allowing species with a common molecular weight to be quantified, for example butan-2,3-dione (MW=86 MH^+=87, glucose degradation product) and 3-methylbutanal (MW=86 MH^+=87, Strecker aldehyde from leucine).

Experimental Procedures

Materials

The following materials were obtained commercially: D-glucose, L-leucine, 3-methylbutanal and α-lactose from Sigma-Aldrich. N-ε-acetyl-L-lysine from Acros Organic. Instant skimmed milk powder from CWS Ltd, Manchester, UK. 250μm silanized glass beads, 250μm acid wash glass beads from Supelco. 1-Amino acetone hydrochloride was a kind gift from Dr Mark Fitzsimons, Petroleum & Environmental Geochemistry Group, University of Plymouth, UK.

Micro-Reactor

A 5.0mm (od) x 3.5mm (id) x 60mm glass tube was co-axially packed with silanized glass beads (apolar surface chemistry, outer annulus) and acid washed glass beads (polar surface chemistry, inner core). The reactor tube was placed in an Optic 2 PTV injector (Atas UK Ltd, Cambridge, UK) set to 20ml/min flow (pressure controlled) and the output flow directed to a Thermo Finnigan LCQ Deca XP ion trap mass spectrometer (Thermo Finnigan Ltd, Hemel Hempstead, UK) using an in-house APcI ion source. A Thermo Finnigan AS800 autosampler was used to inject liquid samples on to the inner core.

Reaction Sequence

The following phases were programmed concurrently using the Optic 2 PTV injector.
Condition phase: The bed was conditioned to remove any contaminants 60°C → 170°C at 1°C/s for 10 min →60°C. *Calibration phase:* Inject 15µl 2.5mM 3-methylbutanal solution onto inner core, allow to dry. *Film generation phase:* Inject 15µl test solution onto inner core, allow to dry. *Reaction phase:* Start water humidifier (water was introduced into the carrier gas stream via a second upstream 'micro-reaction tube' set at 110°C), 60°C for 1 min → 100°C for 5 min → 130°C for 5 min → 170°C for 10 min. Temperature ramp speed was 1°C/s. Mass spectral data was acquired over the range m/z 30-200 for the duration of the reaction sequence. CID ms^2 data on ion 74 was acquired on selected samples.

Color Measurement

The reaction core was removed from the reaction tube and placed in a 2ml vial containing 1ml methanol. Ultrasonication was applied to ensure the complete disruption of the bed and dissolution of the reaction products. Absorbance was determined at 450nm using methanol as reference.

Effect of humidity on volatile and color formation from a glucose-leucine system

A 50mM glucose/50mM leucine solution was injected as the test solution and the 'standard reaction sequence' performed. Five humidity levels were investigated as 0, 0.25, 0.5, 1.0 and 5.0µl/min water. The intermediate humidity level (1.0 µl/min water) was performed in triplicate and used to estimate error for the series. In addition three control samples were performed at humidity level 1.0

μl/min water (50mM glucose only, 50mM leucine only, water). Color measurements were carried out on all samples.

Effect of sugar and amino acid type on color formation

The following systems were injected as the test solution and the 'standard reaction sequence' performed. Solution concentrations were 50mM unless stated. Color measurements were carried out on all samples.
(a) Glucose / 5mM leucine (b) Glucose / N-acetyl-lysine (c) Lactose / leucine (d) Lactose/ N-acetyl-lysine (e) Skimmed Milk Powder/ N-acetyl-lysine.
The skimmed milk powder was added at a level which equated to providing 50mM lactose solution based on a 50% wt/wt lactose content of the dry powder.

Results and Discussions

A schematic diagram and photographic image of the micro-reactor tube show the arrangement of glass beads used to support the liquid film and the form of the reaction core on removal post reaction, respectively (Figure 1).
In preliminary experiments to establish the release characteristics during the formation and drying of the liquid film, a series of aqueous solutions containing different levels of 3-methylbutanal (0.005 -2.5 mM) were injected at 60°C . A typical release profile from this series (0.14mM sample) shows the rapid release of ion 87 (MH^+ , 3-methylbutanal) and ion 37 (MMH^+ , water dimer) (Figure 2). The water dimer maintains a constant plateaued release until the film dries (about 8 min) whilst 3-methylbutanal displays a high initial release which decays rapidly following a logarithmic function (linear regression fit for Log(ion 87) R^2=0.9866) in a time scale preceeding the drying of the liquid film. A release function of this type suggests that release is proportional to the concentration remaining in the liquid film and would be anticipated from thin films. This is a desirable situation as it provides a means to model release and should permit the composite profiles generated to be decoupled into generation kinetics and release kinetics.
Ion profile data from the glucose/leucine series shows a multitude of ion formations as temperature increases through the temperature stages (60°C$\rightarrow$100°C$\rightarrow$130°C$\rightarrow$170°C). At 170°C the primary ions formed for the glucose only sample (1μl/min water) are ion 87, 103, 117, 127, 145 and 163 (Figure 3).

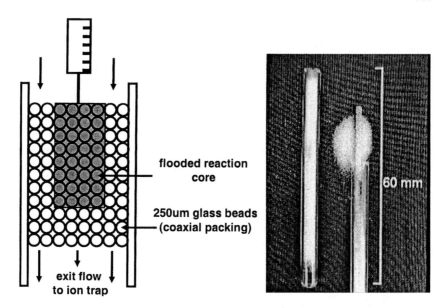

Figure 1. Schematic of micro-reactor tube (right) and photograph of the reaction core post- reaction (left).

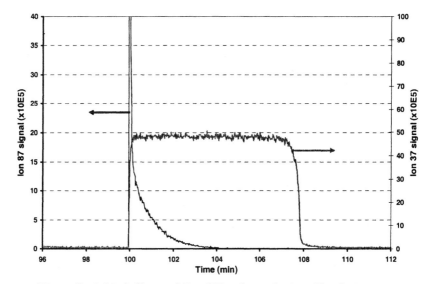

Figure 2. 3-Methylbutanal (ion 87) release during film drying

These can be tentatively assigned from mechanistic projections, based on thermal generation studies (2) and assignment to the molecular ion as butan-2,3-dione (ion 87, diacetyl), 1-hydroxy butan-2,3-dione (ion 103), 4-hydroxy pentan-2,3-dione (ion 117), (5-hydroxymethyl) furfural (ion 127), γ-pyranone or 4-furanone (ion 145), deoxyosone (ion 163). In comparison when the reaction is conducted in the presence of leucine (1 μl/min water) the primary ions formed are ion 74, 88, 142, and 156 (Figure 3).

These ions characteristically show "even" ion masses whereas in the absence of amino acid "odd" ion masses predominate. This position is consistant with ions

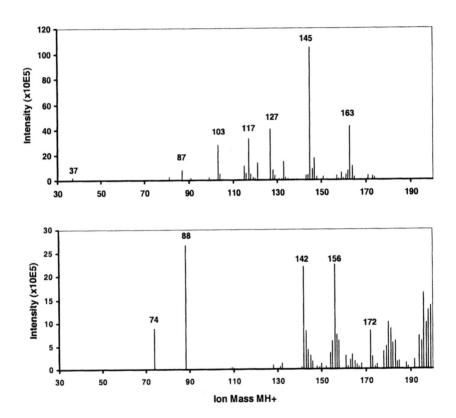

Figure 3. Reaction spectra at 170°C for glucose only (upper) and glucose/leucine (lower)

originating from chemical species containing carbon and oxygen only to yield "odd" ion masses and for chemical species containing nitrogen to yield "even" ion masses. Like before these ions can be tentatively assigned from mechanistic projections as 1-amino propan-2-one (ion 74), 3-amino butan-2-one (ion 88), 1-(3-methyl butylideneamino)prop-1-en-2-ol (ion 142), 3-(3-methyl butylideneamino)but-2-en-2-ol (ion 156). The latter two ions represent intermediates of a classical Strecker degradation mechanism (5) involving 2-oxopropanal (pyruvaldehyde) and butan-2,3-dione (diacetyl) respectively, reacting with leucine through nucleophilic addition to yield these dehydrated, decarboxylate intermediates (Figure 4). The former two ions being transamination products of 2-oxopropanal and butan-2,3-dione respectively.

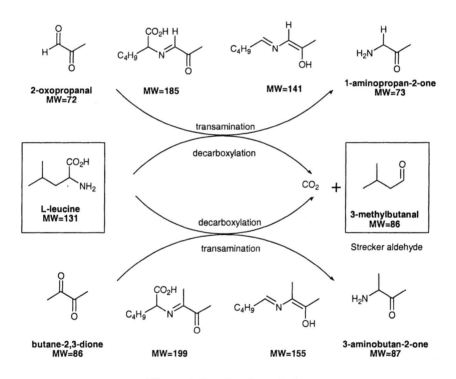

Figure 4. Strecker degradation

To confirm the presence of 1-amino propan-2-one an authentic sample was obtained and a series of aqueous solutions prepared and injected onto the micro-reactor core set at 60°C (data not shown). The ion trap mass spectrometer was configured to perform CID fragmentation on ion 74 using a collision energy of 29%. The CID spectra (ms2) showed two primary ions present, ion 74 (10%, unfragmented parent ion) and ion 56 (85%). Application of commercially available CID fragmentation software (Mass Frontier, High Chem Ltd, UK) predicts the formation of ion 56 from 1-amino propan-2-one protonated at the carbonyl oxygen, via standard fragmentation mechanisms.

The effect of temperature on the formation of ion 74 (ms[1]) and the associated CID ms[2] fragment ion 56 shows the onset of production at 130°C and a further accelerated but transient production at 170°C (Figure 5). The decay in production may be attributed to a number of conditions for example the consumption of precursors, further reaction or perhaps the masking by other species present. The latter two conditions may also account for the apparent absence of ion 87 (3-methylbutanal, Stecker aldehyde) in the spectrum at 170°C (Figure 3, lower spectra) however it is clearly observed at 130°C.

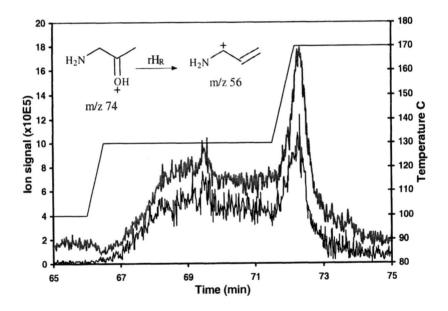

Figure 5   Effect of temperature on the generation of 1-amino propan-2-one from a 50mM glucose/50mM leucine film (humidity at 5µl/min water, ion 74 upper profile, ion 56 lower profile)

The effect of humidity on the formation of 1-amino propan-2-one (ion 74), quantified by the area under the generation profile (Figure 5), shows the rapid production of 1-amino propan-2-one with increasing humidity (Figure 6).

Color determination of the reaction core from the humidity series show an increase in color formation with increasing humidity, however, to a reduced extent compared to 1-amino propan-2-one production (Figure 7). As might be expected the formation of the transamination product of 2-oxopropanal originating from glucose breakdown and the recombination with the amino acid only partially represents those potentially similar chemical reactions which are responsible for color formation.

The effect of amino acid level and type, at constant humidity (1μl/min water), has a significant effect on color formation (Figure 8). Increase in color is observed in the series, glucose < glucose / 5mM leucine < glucose / 50mM leucine < glucose / 50mM N-acetyl lysine. Sugar type also has a significant effect on color formation, increase in color is observed in the series glucose / leucine < lactose / leucine, glucose / N-acetyl lysine < lactose / N-acetyl lysine.

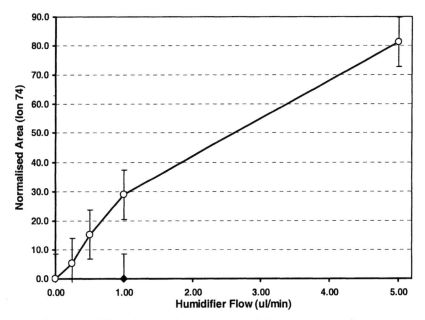

Figure 6. Effect of humidity on the generation of 1-amino propan-2-one from a 50mM glucose/50mM leucine film (black symbol glucose only)

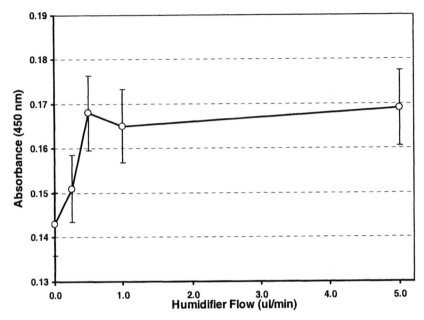

Figure 7 Effect of humidity on the color formation from a 50mM glucose/50mM leucine film

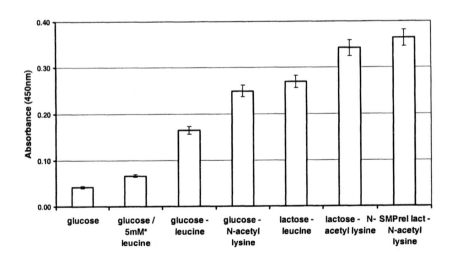

Figure 8. Effect of amino acid and sugar type on color formation from a sugar/amino acid film.

Conclusions

The film reactor-ion trap mass spectrometer technique (FR-ITMS) introduced in this paper offers the potential for novel dynamic investigations of the Maillard reaction. Application of the technique to a glucose/leucine system has shown reactants, intermediates and end products to be successfully monitored in short time scales under rapidly changing reaction environments.

References

1. Tressl, R.; Rewicki, D. In *Flavor Chemistry: 30 Years of Progress*; Teranishi et al., Ed.; ISBN: 0306461994; Kluwer Academic / Plenum Publishers: 1999; pp 305-24.
2. Pischetsrieder, M.; Severin, T. In *Chemical Markers for Processed and Stored Foods*; Lee T-C.; Kim H-J., Ed.; ACS Symposium Series 631; Americam Chemical Society: 1996; pp 14-23.
3. Turner, J.A., Sivasundaram, L.R., Ottenhof, M.-A., Farhat, I.A., Linforth, R.S.T. & Taylor, A.J. (2002). *J. Agric. Food Chem.* 50, 5406-5411.
4. Sivasundaram, L.R., Farhat, I.A. & Taylor, A.J. (2003). In *Handbook of flavor characterization. Sensory analysis, chemistry and physiology.* Eds. Deibler, K.D. & Delwiche, J. Marcel Dekker, New York, pp.379-388.
5. Rizzi, G. P. In *Flavor Chemistry: 30 Years of Progress*; Teranishi et al., Ed.; ISBN: 0306461994; Kluwer Academic / Plenum Publishers: 1999; pp 335-43.

Chapter 16

Formation, Release, and Perception of Taste and Aroma Compounds from Cheeses as a Function of Matrix Properties

C. Salles[1], E. Engel[1], S. Nicklaus[1], A. J. Taylor[2], and J. L. Le Quéré[1]

[1]Unite Mixte de Recherches sur les Arômes, INRA, 17 rue Sully, 21065 Dijon Cédex, France
[2]Samworth Flavor Laboratory, University of Nottingham, Sutton Bonington Campus, Loughborough LE12 5RD, United Kingdom

During cheese ripening, migration phenomena and release of potentially taste-active compounds are responsible for the evolution of sourness, saltiness and bitterness which are the main cheese gustatory characteristics. Moreover, these basic tastes are also influenced by the structure of the matrix that changes during ripening. Correlations between perception parameters evaluated by time-intensity experiments and release of aroma- and taste-active compounds from soft cheeses were also studied, using API-MS (atmospheric pressure ionization mass spectrometry) or HPLC. The matrix and inter-individual differences mainly explained variations observed in taste compound release while the nature of the compound affected only the quantitative release. With aroma release, correlation between temporal perception and flavor release parameters was only found for a sulfury attribute due to sensitivity limitations in the analytical techniques.

Introduction

According to several authors, cheese taste is mainly due to the compounds found in the cheese water-soluble extract (WSE) *(1, 2)*. Thus, to study cheese taste, the focus is usually on the cheese WSE which contains small polar molecules such as minerals, acids, sugars, amino acids, peptides and some volatile compounds produced by different processes such as lipolysis, proteolysis microbial metabolism *(3)*. These compounds are responsible for the individual taste sensations like sourness, bitterness and saltiness which are the main taste descriptors for cheese. However, in a complex mixture they also exert other taste sensations due to taste / taste interactions *(4)*. Peptides are generally considered to be the main bitter stimuli in cheese *(5)*. However, it was shown that in goat cheese, bitterness resulted mainly from the bitterness of calcium and magnesium chlorides, partially masked by sodium chloride *(6)*.

The perception of taste- and aroma-active compounds is modulated by the cheese matrix which is made of water, fat and proteins and undergoes various changes during processing and storage. These changes in matrix structure can lead to changes in release of flavor compounds and consequently on perception. The matrix structure also affects flavor release and perception during mastication of food. The nature of the flavor compounds and the level of mastication parameters *(7, 8)* are also important contributors to flavor release. Specific methods such as atmospheric pressure ionization mass spectrometry (API-MS) were developed for monitoring the release of aroma compounds *(9)* and taste compounds *(10)* in vivo.

The objective of this paper is to bring together original results on flavor release of soft cheese obtained over the last few years. First, results about the evolution of cheese taste during ripening are presented. The effect of the cheese matrix on the evolution of taste-active compounds of the WSE are described. The second part deals with attempts to relate taste- and aroma release parameters with perception parameters in a range of cheeses.

Materials and methods

Chemicals. All synthetic components used in this study were of food grade and were purchased from commercial suppliers. Pure water was obtained from a MilliQ system (Millipore, Bedford, MA).

Experimental cheeses. Experimental French Camembert cheeses (45 % fat dry basis) were manufactured at the "Lycée Agricole de Saint Lô Thère" (Saint Lô, France) as described in *(11)*. A Camembert cheese ripened for 30-days was used for the identification of taste fractions. Camembert cheeses were taken 9, 16, 21, 28, 35 and 42 days after manufacture to study the evolution of taste characteristics and taste-active compounds during ripening. In this case, cheeses were cut into three parts (rind, under-rind and center) after being kept 40 min at -20°C.

Commercial cheese samples. Commercial French cheeses (45 % fat dry basis) were bought at a local supermarket : Camembert « Cœur de lion » made with pasteurized milk, Camembert « Dupont d'Isigny » made with raw milk, and Brie made with pasteurized milk.

Water extraction procedure. The frozen cheese portions were grated, dispersed in pure water (w/w : 1/2) and homogenized for 4 min in a 1094 homogenizer (Tecator, Höganäs, Sweden). The suspension was centrifuged at 20,000 g for 30 min at 4°C. Three phases were separated: a fat upper-layer, a liquid fraction including water-soluble molecules called C20000 and a pellet of proteins. The fat was collected in freezer bags and the protein pellet was dissolved in water corresponding to 7.5 % cheese weight. After filtration on gauze, the C20000 fraction was recovered. All these fractions were stored at −80°C until further use.

Purification procedure. Each C20000 fraction obtained from cheeses ripened for 9, 16, 21, 28, 35 and 42 days was submitted to a 100 kDa frontal ultrafiltration in a cell (V=400mL, d=76mm ; Millipore, Bedford, MA, USA) with regenerated cellulose membranes (Millipore). The filtration temperature and the trans-membrane pressure were maintained at approximately 8°C and 4 Bar respectively. For each fraction corresponding to one of the three cheese portions at one of the six ripening dates, 600 mL of C20000 were treated. The recovered permeate (the water-soluble extract; WSE) was immediately frozen until further use. When used for sensory evaluation, these purified fractions were freeze-dried to compensate for the initial dilution with water due to water-extraction.

The 37.5 kg of C20000 obtained from cheese ripened for 30 days were pooled and submitted to tangential microfiltration in a pilot apparatus equipped with two 0.05 μm membrane modules of 0.9 m² each (INRA, Laboratoire de Recherches de Technologie Laitière, Rennes, France). The retentate was rinsed with 43.5 L of demineralized water corresponding to four successive diafiltrations. Microfiltration permeate (63.2 kg) was considered to be the WSE. All the fractions obtained were frozen at –80°C until further use. When they were used for sensory evaluation, WSE and RUF0.05 were concentrated by freeze-drying to compensate for their dilution with water during the extraction and purification steps.

Preparation of a reconstituted cheese. Starting with the fractions recovered at each step of the extraction / purification process of cheeses, it was possible to prepare a reconstituted cheese if the corresponding fractions - fat, protein pellets, microfiltration permeate (WSE) and retentate - were at the same concentration as in the crude cheese. The dilution of both microfiltration permeate and retentate

(WSE and RUF0.05) were compensated by freeze-drying. Thus, fat, proteins from centrifugation pellets and freeze dried microfiltration permeate and retentate were mixed using an Ultraturrax homogenizer (Polytron 3100, Kinematica, Littau, Switzerland) at 15,000 rpm for15 min.

Omission of fractions. In these cases, reconstituted cheeses were prepared by omitting one of the constitutive fractions. For each omission, the omitted fraction was replaced by ultra-pure water. The mixing of the components was made in the same way as described above. For more details, see *(11)* and *(12)*.

Chemical analysis. Dry mater and pH values were measured on the whole cheese according to *(13)*. For nitrogen determination an aliquot of grated cheese was crushed in a sodium citrate solution, as described in *(14)*. Total nitrogen (TN), soluble nitrogen at pH 4.6 (SN), soluble nitrogen in phosphotungstic acid (SNPTA) were determined by the Kjeldahl method. The results are the average of three replications with a coefficient of variation less than 5%.

Flavor release. The release profiles of key aroma compounds identified by GCO study were monitored by in-nose API-MS *(9)* and the changes in sensory attributes over the course of eating were evaluated by time intensity measurements. As soon as the panelists introduced a sample of cheese into their mouths, their breath was guided from one of their nostrils to the interface of the mass spectrometer where ions were monitored. Simultaneously the panelists were asked to rate the intensity perception of one sensory attribute on a marked scale while mastication of the sample was taking place. The procedure is detailed in *(15)*. To study non-volatile compound release, samples were swabbed from the tongues of the panelists using cotton buds during 3 min of mastication of cheese (approximately one swab each 10 sec) *(10)*. The swabs were extracted, analyzed by direct API-MS to build corresponding tastant release curves.

Sensory evaluation. The procedure used for the evaluation of samples of Camembert cheese fraction prepared by the omission method and the individual components are described in *(11)* and *(15)* respectively. For each perceptual time intensity (TI) curve obtained, primary and secondary parameters were calculated for a trapezoid model drawn between a point representing the time corresponding to 5 % and 90 % of the intensity scale in the increasing and the decreasing phase of TI curves. Secondary parameters were calculated from these primary ones : the duration D, the rate R and the area A in the increasing, decreasing and middle parts of the curve. For flavor release parameters, analogous calculations were made with the primary parameters corresponding to 25, 75, 90 % of the intensity scale.

Statistical treatments. The data were processed with the SAS statistical package version 6.11, 4th edition (SAS Institute, Inc., Cary, NC). ANOVA analyses were performed at level $\alpha = 0.05$, according to the model attribute = product + subject + product x subject, with subject as a random effect. Means were compared with the Newman – Keuls multiple comparison test (Student *t* test). To quantify the relative impact of each compound on each attribute, stepwise multiple linear regressions were performed with proc REG with the stepwise option to select the variables. For flavor release and temporal perception studies, ANOVA was carried out with SPSS version 10.1. Principal component analysis of the perception and quantitative mass spectrometry data were drawn with Guideline + 7.2 (Camo, Trondheim, Norway).

Results and discussion

Importance of cheese fractions on taste perception

For this study, an experimental, strongly bitter Camembert cheese was ripened for 30 days. Its pH was 5.94 and it contained 9.45 g soluble nitrogen / kg cheese and 7.05 g peptides / kg cheese. The methodology to determine cheese taste compounds is shown in Figure 1. After extracting water soluble extract containing taste compounds by grinding cheese in water and separating lipids and proteins, the WSE composition was determined. A model solution was made with authentic compounds apart from peptide fractions which were prepared by filtration of the WSE. The contribution of each fraction and component of the WSE to taste was determined by omission tests made with the synthetic WSE. These tests consisted in omitting one or several elements of the solution and in comparing the taste characteristics and intensity to determine taste activity (12).

Table 1 shows the taste profile obtained for crude cheese, reconstituted cheese made with homogenized proteins, fat with and without WSE, and WSE. Their comparison allowed the impact of each fraction on the taste of the cheese to be evaluated. The omission of WSE led to a tasteless product, showing that WSE contained all the taste-active compounds. In reconstituted cheese where the structure of the matrix was almost totally degraded, bitterness was weaker and saltiness higher than in crude cheese (Table 2). The omission of fat and proteins from the reconstituted cheese caused an increase of saltiness and a decrease of bitterness compared to crude cheese. These data demonstrated that, in the crude cheese, the matrix structure partially masked the saltiness and increased the bitterness due to taste-active compounds. In addition, the comparison with results obtained with grated cheese in which the destructuring was intermediate between crude and reconstituted cheese for the same taste descriptors (Table 2) confirmed that the more the matrix was destructured, the more the bitterness increased and the saltiness decreased. Thus, cheese taste might be explained by the taste of the WSE containing the taste-active compounds modulated by the masking effect of both fat and proteins but also by an effect linked to the cheese matrix structure.

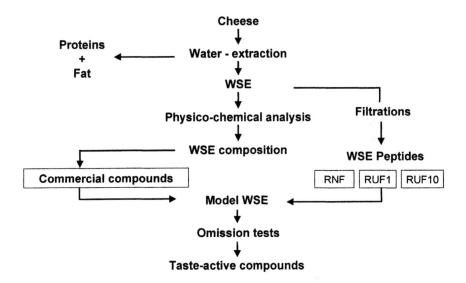

WSE : water-soluble extract ; RNF, RUF and RUF10 are the retentates obtained respectively by ultrafiltration of the WSE on 10 KDa, 1 KDa and nanofiltration on a 0.5 KDa molecular weight cut-off membranes.

Figure 1. Methodology for cheese taste study

Table 1. Comparison between gustatory profile of reconstituted cheese, reconstituted cheese where fat and proteins were omitted and reconstituted cheese where the Water soluble extract (WSE) was omitted.

Product	Sour	Bitter	Salty
Reconstituted Cheese	21.8[a] *(20)*	35.4[a] *(23)*	65.8[b] *(18)*
RC – Fat - Proteins	19.2[a] *(21)*	35.7[a] *(28)*	76.5[b] *(16)*
RC - WSE	1.7[b] *(4)*	4.5[b] *(9)*	1.7[c] *(4)*

For each attribute, the means with the same letter (a, b, c) are not significantly different at the level of 5% according to Newman-Keuls tests. All omissions have been performed for two replicates (16 panelists). Standard deviations are indicated in italic between brackets.

Table 2. Gustatory profile of crude, grated and reconstituted cheese.

Product	Sour	Bitter	Salty
Entire cheese	23.4 *(27)*	73.2 *(22)*	46.3 *(27)*
Grated cheese	22.5[ns] *(24)*	62.6[a] *(21)*	60.1[b] *(30)*
Reconstituted cheese	21.8[ns] *(20)*	35.4 *** *(23)*	65.8** *(18)*

For each attribute, t-tests were performed between entire cheese and grated or reconstituted cheese. Significant ($P<0.05$), very significant ($P<0.01$) and highly significant differences compared to the entire cheese are indicated with *, ** or *** respectively. When the P-value is higher than 0.05 but weaker than 0.10 it is shown between parenthesis. ns = no significant. Each product was tasted twice by each of the 16 panelists. [a]$P<0.09$; [b]$P<0.1$. Standard deviations are indicated in italic between brackets.

Consequently, we focused our study on WSE. The taste profile of model WSE where different fractions of peptides (RNF, RUF1, RUF10) were omitted, showed that bitterness was mainly due to peptides, and particularly to the RUF1 and the RNF fraction containing peptides with a MW lower than 10 kDa (see *12*). We paid particular attention to these two bitter fractions. RP-HPLC-tandem mass spectrometry led us to identify 40 peptides. Among them, 38 peptides were hydrophobic enough to be potentially bitter. Most of them were related to particular zones of casein (alpha S1 and beta), with a molecular weight between 900 and 2100 (Results not shown).

The results of omission tests (*12*) including other compounds in WSE and attributes are reported in Table 3. Cheese sourness was explained by the enhancing effect of sodium chloride on the sourness due to hydronium ions concentration. Cheese bitterness was explained by the enhancing matrix effect on bitterness due to small peptides. Cheese saltiness was respectively explained by a partially masking effect of the matrix on the WSE salty taste due to sodium chloride.

Table 3. Omission tests results made on WSE constituents.

Cheese taste	WSE taste	Matrix effect
Sourness	H_3O^+ / NaCl	
Bitterness	Peptides (< 10 Kda)	Enhancing effect
Saltiness	NaCl	Masking effect

WSE : water-soluble extract

Evolution of taste characteristics during cheese ripening

The effect of the matrix on the development of taste characteristics was then studied. It is well known that the cheese matrix is degraded during ripening mainly because of proteolysis and lipolysis. Adequate time must be allowed for cheese taste to fully develop. Cheeses of different types were studied at six different ripening stages. The three portions : rind, under-rind and center, were separately investigated. For each fraction at each ripening stage, sensory and physicochemical analyses were conducted for the crude cheese fractions and the corresponding WSE to measure taste intensity data for each taste descriptor as well as the concentration of the taste-active compounds. However, as previously shown, a matrix effect had to be taken into account to fully explain the perceived taste of the cheese. This matrix effect can be demonstrated for bitterness where the bitterness intensity perceived in cheese decreased during ripening whereas bitterness intensity perceived in the corresponding WSE, increased during ripening (Figure 2). It implied that the increase in cheese bitterness might be partly explained by an increase in the enhancing effect of matrix on WSE bitterness. Thus, the matrix degradation due to proteolysis and lipolysis might increase its enhancing effect on the bitter taste of WSE.

There was no effect of the matrix on sourness or saltiness, the destructuring of the matrix led to a slight decrease in the matrix effect. That could be easily interpreted as a decrease of the masking effect of the matrix as it becomes destructured during ripening.

The explanation of the matrix effect on bitterness was unclear. Overall, the measured matrix effect (Figure 2) increased along the ripening time. That indicated bitterness was significantly more intense in WSE than in cheese. This result confirmed previous observations made from omission tests with cheese fractions (11). Phenomenon such as lipolysis and proteolysis, leading to a modification of the matrix structure, could be responsible for taste exhausting effect observed during cheese maturation. That means the origin of matrix effect could be due to texture/taste interactions.

However, we can not exclude direct modifications of the stimuli. As shown in Figure 3, the peptide composition of the cheese seemed to evolve during ripening. So, it could be hypothesized that peptides synthesized at the end of

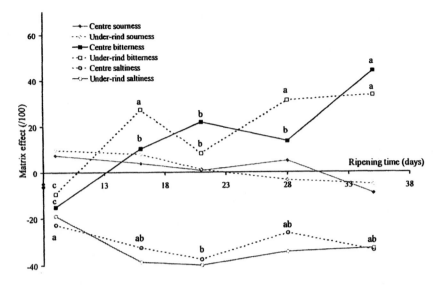

Matrix effect on a taste X is defined as the difference between the mean perceived intensity of X in the cheese and the mean intensity of X in the corresponding water-soluble extract. For each attribute, the means with the same letter (a, b, c) are not significantly different at the level of 5% according to Newman-Keuls tests. Calculation of each matrix effect value was performed starting with two replications of taste evaluation for both cheese and water-soluble extract (14 panelists). Reproduced from reference 11. Copyright 2001 American Chemical Society

Figure 2. Evolution of the matrix effect on sourness, bitterness and saltiness during ripening of a 30-days ripening camembert cheese.

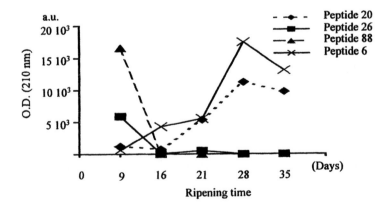

Figure 3. Example of peptide evolution during ripening in the under-rind of Camembert cheese. O.D.: optical density ; a.u.: absorbency unit.

maturation interacted differently with the matrix leading to different effect of the matrix compared to the beginning of ripening. That could at least explain the increase of matrix effect.

Release of flavor compounds during chewing

This part concerned attempts to relate taste- and aroma release parameters with perception parameters from a complex matrix such as soft cheeses.

For aroma as for taste analysis, the approach was first to compile a sensory profile by frequency of citation. The 3 soft cheeses used (Brie and Camembert made with pasteurized milk, respectively BP and CP, and Camembert made with raw milk, CU) could be described by the same main descriptors : sulfury, buttery, mushroomy, salty and sour but some differences in frequency citation were observed, in particular for the sulfury note which was higher in the 2 Camembert cheeses.

Time-intensity measurement were made for these attributes and flavor release was evaluated for the main detectable volatile and non volatile compounds.

Comparison of release profiles of 7 amino acids for 2 panelists for CU is presented in Figure 4. Its seemed that for each panelist, though different levels of concentration are observed, the general patterns seem rather similar and time parameters were not significantly different.

If we considered inter-individual aspects, different release patterns were observed according to the panelist, for the general pattern of the curves and time parameters. The quantity of compound released by panelist 1 was higher than panelist 2. Panelist 1 needed 2 min to release the maximum concentration of compound while panelist 2 achieved it in less than 1 min. Those inter-individual differences were probably significant for perception variability and were probably linked to physiological parameters of chewing. These results agreed with the findings of Pionnier et al. *(7, 8)*.

Some comparison of release curves for different compounds from the three cheeses are shown in Figure 5. Though important differences were observed in the quantity of compound released, between the 3 cheeses, we can observe that the profiles for the same compound and the same panelist are rather different according to the cheeses. That indicated the composition of the matrix, different from one cheese to another, was also a significant factor of variation for non volatile compound release.

Concerning volatile compounds, although API-MS sensitivity was in the ppb range and we identified 19 odor-active compounds by GCO, nose-space experiments only detected six ions at detectable limits during eating of these cheeses. These ions corresponded namely dimethylsulfide (m/z 63), S-

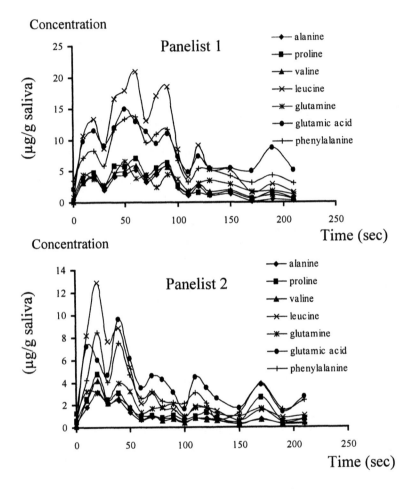

*Figure 4. Release of amino-acids from soft cheeses during mastication :
examples with Camembert cheese made with raw milk.*

methylthioacetate (m/z 91), dimethyldisulfide (m/z 94), 3-methylbutanal / diacetyl (m/z 87), 2-heptanone (m/z 115) and 2-nonanone (m/z 143) which were found to be odor-active.

The most significant parameters were selected by ANOVA analysis (not shown). Correlation, mainly for the sulfury note, between the perception parameters derived from the TI curves and parameters derived from the aroma release curves are presented in Figure 6. The sensory characteristics of the sulfury note were denoted by the initial letter (S) while the aroma release of individual compounds is denoted by a number which relates to the m/z value. PC1 seemed to represent the sulfury characteristic axis. Time and area parameters, which were related to the most intense parts of the TI curves, were represented on the positive part of PC1 while the decrease rates of sulphur components were on the opposite end of the axis. PC2 seemed representative of methylketone (the ions at m/z 143 and 115 are 2-nonanone and 2-heptanone respectively) but they did not relate to any perception parameter.

No or very low correlations were obtained for buttery and mushroom notes, and related ions. The main reason could be the limitation due to API sensitivity and selectivity but in some case, the complex volatile composition necessary to explain a particular note could explain this lack of correlation.

Conclusion

In cheese, a matrix effect could modulate salty and bitter perception in different ways and evolve during the maturation process. The composition of the matrix and physiological parameters had to be taken into account to better understand temporal release and perception of flavor. The study of the relationships between sensory and aroma release gave reliable results only for sulfury note which was the most intense and due to well identified and specific compounds.

The challenge of our further studies is to elucidate matrix effect for bitterness and to try to relate salty and sour perception of soft cheeses with mineral and organic non volatile release parameters.

References

1. Aston, J. W.; Creamer, L. K. *N. Z. J. Dairy Sci. Technol.* **1986**, *21*, 229-248.
2. Biede, S. L.; Hammond, E. G. *J. Dairy Sci.* **1979**, *62*, 238-248.
3. Sablé, S., Cottenceau G. *J. Agric. Food Chem.* **1999**, *47*, 4825-4836.

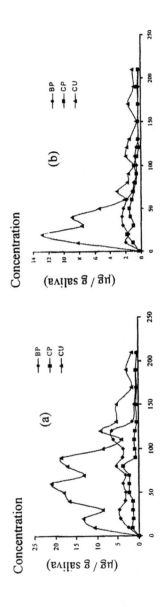

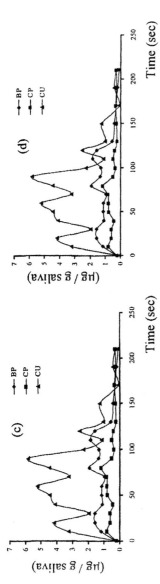

BP : Brie cheese made with pasteurized milk ; CP : Camembert cheese made with pasteurized cheese ; CU : Camembert made with raw milk
(a) L-leucine / panelist1 ; (b) L-leucine / panelist 2 ; (c) L-alanine/ panelist 1 ;
(d) L-glutamic acid / panelist 1

Figure 5. Release of amino-acids from soft cheeses during mastication : comparison between the three soft cheeses.

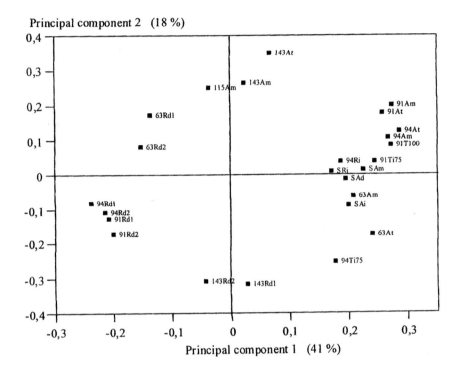

Numbers represent the m/z values of the compounds, S represent the sensory score, the following descriptors denote the parameters derived from analysis of the aroma release and time intensity curves (see Experimental for details). Reproduced from reference 15. Copyright 2003 Lavoisier

Figure 6. Principal component analysis scores for the sulfury attributes of the three soft cheeses.

4. Keast, R.S.J.; Breslin, P.A.S. *Food Qual. Pref.* **2002**, *14*, 111-124.
5. Habibi-Najafi, M. B.; Lee, B. H. *Critical Reviews Food Sci. Nut.* **1996**, *36*, 397-411.
6. Engel, E., Nicklaus, S., Septier, C., Salles, C., Le Quere, J. L. *J. Agric. Food Chem.* **2000**, *48*, 4260-4267.
7. Pionnier, E., Chabanet, C., Mioche, L., Le Quéré, J. L., Salles, C. *Journal of Agricultural and Food Chemistry* **2004**, *52*, 557-564.
8. Pionnier, E., Chabanet, C., Mioche, L., Taylor, A. J., Le Quéré, J. L., Salles, C. *Journal of Agricultural and Food Chemistry* **2004**, *52*, 565-571.
9. Taylor, A. J., Linforth, R.S.T., Harvey, B.A., Blake, A. *Food Chem.* **2000**, *71*, 327-338.
10. Davidson, J..M., Linforth, R.S.T., Hollowood, T., Taylor, A. In *Flavor release*; Robert, D. & Taylor, A., Eds., American Chemical Society: Washington, D.C., 2000; pp 99-111.
11. Engel, E., Nicklaus, S., Septier, C., Salles, C., Le Quere, J. L. *J. Agric. Food Chem.* **2001**, *49*, 2930-2939.
12. Engel, E., Septier, C., Leconte, N., Salles, C., Le Quere, J. L. *J. Dairy res.* **2001**, *68*, 675-688.
13. Molimard, P.; Lesschaeve, I.; Bouvier, I.; Vassal, L.; Schlich, P.; Issanchou, S.; Spinnler, H. E. *Lait* **1994**, *74*, 361-374.
14. Gripon, J. C.; Desmazeaud, M. J.; Le bars, D.; Bergère, J. L. *Lait* **1975**, *55*, 502-516.
15. Salles, C., Hollowood, T., Linforth, R. S. T., Taylor, A. J. In *Flavour research at the dawn of the 21st century*; Le Quéré J. L. & Etiévant, P., Eds, Lavoisier, Paris, 2003; pp 170-175.

Chapter 17

Nature, Cause, and Control of Irradiation-Induced Off-Odor in Ready-to-Eat Meat Products

Xuetong Fan

Food Safety Intervention Technologies Research Unit, Eastern Regional Research Center, Agricultural Research Service, U.S. Department of Agriculture, Wyndmoor, PA 19038

Ionizing radiation improves food safety and extends shelf life by inactivating food-borne pathogens and spoilage microorganisms commonly found in ready-to-eat (RTE) meat products. However, irradiation may induce the development of an off-odor, particularly at high doses. The off-odor has been called "irradiation odor" and described as "sulfide", "wet dog" and "barbecued corn-like". Although the exact compounds responsible for the off-odor are not completely clear, substantial evidences suggest that volatile sulfur compounds (VSCs) play an important role in the development of the off-odor. Many sulfur compounds, induced by irradiation, have low odor thresholds. These compounds include hydrogen sulfide, methanethiol (MT), methyl sulfide, dimethyl disulfide and dimethyl trisulfide. In comparison, thermal processing (heating and microwave) mainly induced MT and ethyl methyl sulfide. VSCs were presumably synthesized from sulfur containing compounds (such as methionine, cysteine, thiamine, glutathione) reacting with free radicals generated from water radiolysis. Antioxidants applied either as ingredients in raw meat emulsions prior to RTE manufacture or as post-manufacture dipping did not consistently reduce VSCs formation caused by irradiation. Research is needed to explore means of controlling the production of VSCs and off-odor in irradiated RTE meat products.

Introduction

Consumers demand convenient, fresh, natural and healthy ready-to-eat (RTE) foods as results of changes in eating habits, family size and lifestyle (*1*). However, many RTE products including prepared sandwiches, bologna, hams, frankfurters, meat spreads, meat salads, and refrigerated RTE meats have been recalled and/or associated with outbreaks of foodborne illness due to contamination of human pathogens, mainly *Listeria monocytogenes*, *Salmonella* and *E. coli* O157:H7 (*2*, *3*). *L. monocytogenes* is capable of growth at refrigerated temperatures, and the mortality rate associated with listeriosis is high for the 'at-risk' populations, such as elderly, newborns and immune-compromised. Consequently, the FDA and USDA have adopted a "zero tolerance" policy regarding the presence of *L. monocytogenes* in RTE products, requesting processors and federal establishments to take meaningful steps to reduce the incidence of *L. monocytogenes.*

Manufacture of RTE meat products involves a cooking step which should eliminate the common foodborne pathogens. It is believed the pathogens that cause the illnesses/recalls of RTE meat products are a result of post-pasteurization contaminated during cooling, rinsing, slicing, and packaging. Ionizing radiation, a nonthermal processing technology, effectively inactivates food-borne pathogens, reduces spoilage and extends shelf life of both raw and processed meat products (*4*, *5*, *6*). When used as a terminal step (i.e. used after packaging), irradiation can help to avoid microbial problems associated with post-pasteurization contamination. U.S. FDA permits use of irradiation only in raw meats including red meats, poultry, and pork. A petition has been filed to the FDA requesting the use of irradiation at doses up to 4.5 kGy in RTE products (*7*).

Despite the obvious benefits of irradiation, commercial use of the technology is limited due partially to concerns about adverse effects of irradiation on product quality. In addition, despite the completion of hundreds toxicology studies showing chemical safety of irradiated products over last 45 years, some anti-technology groups still question the safety of irradiated foods. One of the concerns is the development of an off-odor. A cooking step, applied prior to consumption of irradiated raw meats, eliminates much of the irradiation-induced color changes and off-odor. Hashim et al. (*8*) reported that the off-odor due to irradiation was non-detectable after irradiated chicken was cooked. Heating after irradiation also reduced the amount of volatile sulfur compounds (VSCs) in RTE product (*6*). However, RTE foods are routinely consumed without the cooking process even though heating is recommended for some RTE products due to microbial safety concerns. Therefore, the off-odor may be a particular concern for RTE meats. Over the years, several excellent review articles have been published covering some aspects of irradiation-induced off-odor and VSCs, mainly in raw meats (9, 10, 11). In the present article, the nature of off-odor induced by irradiation, the involvement of VSCs in the off-odor, control of off-odor, and the need for future research will be discussed. A comparison study of irradiation and thermal processing effects on VSCs will also be presented.

The Nature of Off-Odor

Volatile compounds are an integrated part of food flavor. When foods are irradiated, particularly at high doses, an off-flavor can develop. The off-flavor/off-odor has been called "irradiation odor" and described as 'metallic', 'sulfide', 'wet dog', 'wet grain' (12, 13). There are only a few studies on the off-odor development in RTE products. Most of the studies have been focused on raw meats. When RTE beef and pork frankfurters were irradiated at doses of 8 and 32 kGy (irradiation temperature: -34°C), off-odor and off-flavor were noticed and the intensity of off odor increased with radiation dose (14). Barbut et al. (15) found three of the four irradiated frankfurter (5 and 10 kGy, –30°C) were scored significantly higher in off-flavor than the non-irradiated ones. Al-Bachir and Mehio (16) found RTE beef luncheon meats irradiated at doses of 2-4 kGy had no difference in off-flavor compared to the non-irradiated controls. Recently, Johnson et al. (17) showed that the aroma of cooked diced chicken meats and chicken frankfurters irradiated at doses up to 3 kGy (irradiation temperature: 4°C) did not differ from the non-irradiated ones. After 18 days of storage, the aroma of irradiated diced chicken was better than the control presumably due to inactivation of spoilage microorganisms by irradiation. Descriptive analysis of the-off odor was reported mostly on irradiated raw meats. Hashim and others (8) reported that irradiated uncooked chicken thigh had a higher 'blood and sweet aroma' than non-irradiated, Heath and others (18) reported that the odor of irradiated uncooked chicken breast and thigh produced 'hot fat', 'burned oil' and 'burned feathers' odors. Ahn et al. (19) described the off-odor as 'barbecued corn-like'. Based on the limited number of reports on RTE meat products, it is obvious that high dose radiation can cause an off-odor, even if the products are irradiated at frozen state.

Involvement of volatile sulfur compounds in irradiation odor

Evidence indicates that VSCs are most responsible for the off-odor due to irradiation. This evidence includes: 1) The irradiation odor is different from rancidity, which is believed to be caused mainly by lipid oxidation. 2)

Irradiation of the lipid (fat soluble) phase of a meat extract does not produce the characteristic off-odor while irradiation of the aqueous (water soluble) portion of the meat extract results in a typical irradiation odor (20). 3) Irradiation of sulfur-containing amino acids or polypeptides produced a similar off-odor as the irradiation odor (21). 4) The amount of VSCs increased with radiation dose while volatiles from lipids were not always correlated with radiation dose (19). Several earlier researchers suggested that hydrogen sulfide (H_2S) and methanethiol (MT) were important for the development of the off-odor (12, 20, 22). Patterson and Stevenson (23), using GC-olfactory analysis, showed that dimethyl trisulfide (DMTS) was the most potent off-odor compound in irradiated raw chicken meats followed by cis-3- and trans-6-nonenals, oct-1-en-3-one and bis(methylthio-)methane. Ahn and his colleagues have published extensively on irradiation-induced volatile compounds in raw meats (11). They have identified MT, dimethyl sulfide (DMS), dimethyl disulfide (DMDS) and DMTS in different types of irradiated raw meats using GC-FID and GC-MS.

There were limited studies on VSCs in RTE meat products. Du and Ahn (24) found that irradiation induced formation of MT, DMDS and DMTS in turkey sausage. The low levels and reactivity of volatile sulfur compounds complicated accurate detection of these compounds. Recently, we have used a solid-phase microextraction (SPME) technique and a pulsed flame photometric detector (PFPD) for VSCs. PFPD is very sensitive to sulfur compounds, detecting VSCs in part per trillion (ppt) ranges. Use of the SPME technique avoids the formation of artifacts due to high temperature as used in many other extraction techniques, however, SPME techniques have low repeatability, resulting in larger variations among replicates. Fan et al. (2002) identified SO_2, H_2S, MT, CS_2, DMS, DMDS and DMTS in irradiated cooked turkey breasts using SPME-GC-PFPD. Most of the VSCs were promoted by irradiation in a dose dependent manner in the RTE turkey meat. CS_2 levels were however, reduced by irradiation. H_2S and MT decreased rapidly during storage at 4°C even under air-impermeable vacuum packaging (25, 26). The disappearance of the low-boiling-point sulfur compounds may be due to their reactivity and instability. Figure 1 illustrates irradiation-induced VSCs in commercial RTE turkey bologna. Six VSCs were identified, including H_2S, CS_2, MT, DMS, DMDS and DMTS. Irradiation increased levels of MT, DMDS, and DMTS, did not affect the levels of H_2S, and decreased CS_2 and DMS levels in the turkey bologna (Fig. 1). It appears that irradiation can either increase or decrease the levels of H_2S or DMS depending on meat composition, initial concentration of the compounds, packaging type, and atmosphere composition (data not shown). Many of the VSCs are highly reactive and unstable. For example, H_2S in aqueous solution becomes elemental sulfur upon reacting with oxygen while DMDS converts to DMS and DMTS (Fig. 2). Our studies show that MT, DMDS and DMTS are the only VSCs that are consistently promoted by irradiation in RTE meat samples.

Many of the VSCs have very low odor thresholds (in ppb and ppt ranges). Table I lists odor thresholds of selected VSCs in water and in air. MT appears to be the most potent compound with an odor threshold of 0.02 and 0.5 ppb in water and air, respectively. Our results showed that MT was most induced by irradiation, relative to the control (25, 26). Therefore, MT is likely involved in the development of the off-odor.

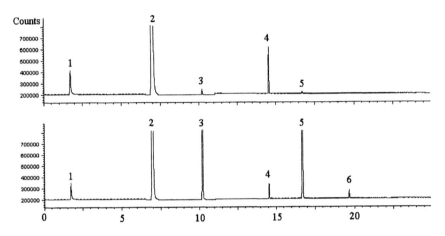

Figure 1-. Chromatogram of volatile compounds in non-irradiated (top) and 3 kGy-irradiated (bottom) commercial RTE turkey bologna. VSCs in the diced turkey bologna were measured using solid phase microextraction and GC-pulsed flamed photometric detector. The volatile compounds are: 1= hydrogen sulfide, 2=carbon disulfide, 3=methanethiol, 4=dimethyl sulfide, 5=dimethyl disulfide, 6=dimethyl trisulfide.

Table I. Odor Thresholds of Selective VSCs in Water and Air

Compounds	Threshold concentration (ppb)	
	Water	*Air*
Hydrogen sulfide	5	0.1-180
Methanethiol	0.02	0.5
Dimethyl sulfide	0.3-10	2-30
Dimethyl disulfide	0.2-50	3-14
Dimethyl trisulfide		7.3
Carbon disulfide		70
Sulfur dioxide		5000

Adopted from Shankaranarayana et al. (27); and Van Gemert and Nettenbreijer (28).

An odor is not only impacted by the concentration of individual compounds, but also by the relative balance (ratio) among volatile compounds, and by semi- and non-volatile compounds. It is likely that other non-sulfur compounds are also involved in the development of irradiation odor. The possible non-sulfur volatile contributors to the irradiation odor may include acetaldehyde, acetone, ethanol, methanol, methyl ethyl ketone, nonanal and phenylacetaldehyde (9). A mixture of 3-(methylthio)propionaldehyde (methional), nonanal and phenylacetaldehyde has been shown to have a similar odor as irradiation odor in beef (22, 29).

Upon irradiation of water at 25°C, the following reaction occurs: $H_2O \rightarrow e_{aq}^-$ (2.8) + H_3O^+ (2.8) + $\cdot OH$ (2.8) + $\cdot H$ (0.5) + H_2 (0.4) + H_2O_2 (0.8). The numbers in parenthesis are the relative amounts expressed as G-values (number of species per 100 eV absorbed) (30). The primary free radicals generated from radiolysis of water are hydrated electron (e_{aq}^-), hydroxyl radicals ($\cdot OH$) and hydrogen atoms ($\cdot H$). The VSCs found in irradiated meat products are likely formed from sulfur containing compounds reacting with the free radicals generated from the radiolysis of water. These sulfur containing compounds may include amino acids in the form of either free amino acids (methionine, cysteine), peptides (glutathione and cystine) or proteins, and others (thiamine, coenzyme A). Figure 2 shows a proposed formation of DMS, DMDS and DMTS as a result of methionine reacting with hydrated electrons. Many VSCs may be produced by secondary reactions of the primary VSCs.

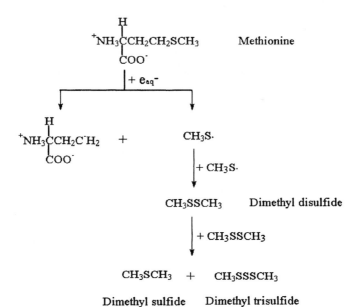

Figure 2. Proposed formation of methyl sulfide, dimethyl sulfide and dimethyl disulfide from methionine (adopted from Yoo et al.(31) with permission).

214

Martin et al. (*32*) suggest that most of MT formed by irradiation is directly from methionine, though some was produced during irradiation of glutathione, possibly indirectly. Although H$_2$S was found in irradiation of both methionine and glutathione, most H$_2$S originated from other sulfur-containing precursors in meats (*32*). In the absence of oxygen, major products arising from irradiation of cysteine solution at pH 5-6 are cystine, alanine, H$_2$S and hydrogen (*33*). In the presence of oxygen, the products are cystine and hydrogen peroxide (*34*). Irradiation of oil containing methionine produced MT, DMDS, DMTS and 3-methylthiopropanal (*35*) while no DMDS was found during the irradiation of a polypeptide containing methionine (*21*), indicating that side chains and portions other than sulfur moieties of sulfur containing compounds also influence the formation of volatile sulfur compounds. Irradiation of glutathione produced CS$_2$ and DMDS (*21*). Ahn's group identified many VSCs using GC-mass spectrometry from irradiated amino acid polymers (polypeptides) in model systems (*11*). Most of the compounds found in the radiolysis of sulfur-containing compounds were not routinely found in meat samples. Methionine appears to be the most radiation-sensitive sulfur containing compound, and the major source of many VSCs. Table II lists some possible sources of common VSCs.

Table II. Major VSCs Produced by Irradiation of Sulfur-containing Amino Acids

Amino acid	Sulfur compounds produced
Cysteine	Sulfur dioxide, hydrogen sulfide, carbon disulfide, dimethyl disulfide, methanethiol
Cystine	Carbonyl sulfide, carbon disulfide
Methionine	Methanethiol, dimethyl sulfide, dimethyl disulfide
Glutathione	Carbon disulfide, dimethyl disulfide

Adopted from Merritt Jr. (*36*), Ahn (*21*), Ahn and Lee (*11*), and unpublished data.

Comparison of VSCs Induced by Thermal Processing or Irradiation

A study was performed to compare VSCs induced by thermal processing or irradiation of RTE turkey bologna. Turkey bologna, purchased from a local supermarket, was diced and placed into 40 ml vials (10 grams of sample in each vial). The vials were then sealed using Teflon-lined septa. The samples were then subjected to the following treatments: 1), None (4°C); 2), 95°C waterbath for 10 min; 3), microwave for 10 sec (temperature reached 90-100°C at the end of treatment); 4), 2 kGy gamma radiation at 4°C. After treatment, the samples were stored at 4°C overnight. VSCs were then extracted using the SPME technique, and analyzed using a GC coupled with a PFPD. The procedures and conditions for the SPME-GC-PFPD were similar to our previous study (26) except that the column length for the GC was shorter (8 m) in this study. The experiment was repeated five times conducted in different days using different packages of bologna. Table III lists VSCs induced by thermal processing or irradiation. Relative amounts of VSCs are expressed as square root of peak area because the sulfur response of PFPD is purely quadratic (37). Gamma irradiation induced mainly MT, DMDS and DMTS and reduced H_2S, CS_2 and DMS. The 95°C treatment increased MT levels and CS_2 levels while microwave radiation induced MT and ethyl methyl sulfide. It is unclear why only the microwave radiation induced EMS and the 95°C treatment did not, although both are thermal treatments. An earlier study has shown that high-temperature (121-131°C) treatment induced formation of H_2S in ground pork meat (38). We did not observe the increase in H_2S perhaps due to the lower temperature we used for the thermal treatments. Our results indicated that thermal treatment (particularly microwave) induced production of some VSCs although not necessarily the same VSCs as those induced by gamma irradiation.

Table III. Effect of Irradiation and Thermal Processing on Volatile Sulfur Compound Production of Turkey Bologna

Treatment	H_2S^Z	CS_2	MT	DMS	EMS	DMDS	DMTS
None	458 a	6908 b	654 c	480 a	0 b	176 b	65 b
95C	520 a	7674 a	1179 b	523 a	0 b	179 b	68 b
Microwave	451 a	7137 ab	1166 b	479 a	514 a	193 b	68 b
Gamma	229 b	5064 c	3741 a	277 b	22 b	1486 a	176 a
$LSD_{0.05}^Y$	86	572	123	55	63	49	10

Diced turkey bologna was either not-treated (none) or treated at 95°C for 10 min, microwaved for 10 sec or gamma irradiated at 2 kGy. Volatile compounds were measured the day after the treatments, and expressed as square root of peak area count. MT=methanethiol. DMS=dimethyl sulfide, DMDS=dimethyl disulfide, EMS=ethyl methyl sulfide, DMTS=dimethyl disulfide. The numbers are means of five replicates
[Z] Numbers in column followed by the same letter are not significantly different (P<0.05).
[Y] The least significant difference at P<0.05 level.

Reduction of Volatile Sulfur Compounds and Off-Odor

Many factors influence the development of irradiated odor. The primary factors appear to be irradiation dosage, the temperature of irradiation, use of food additives and atmosphere (oxygen). Huber et al. (12) suggested three techniques to reduce off-odor: irradiation in the frozen state, irradiation in the absence of oxygen, and irradiation in the presence of antioxidants, such as gallic acid, tocopherol or in combination with citric acid or ascorbic acid. Meats packaged in helium appear to have better taste (39). The composition of foods, storage temperature, atmosphere, and other factors may also have an effect. Generally RTE meats are stored and displayed in refrigeration temperatures without prior exposure to frozen temperatures. Therefore, the irradiation of foods in a frozen state is not applicable to RTE meats. Most of the research has been focused on using antioxidants and packaging systems to control off-odor in raw and RTE meats.

It has been known that various antioxidants reduce VSCs in raw meats. Nam et al. (40) showed that addition of antioxidants such as tocopherols, gallic acid and sesamol reduced the production of some VSCs in raw pork homogenates and patties. Addition of ascobic acid at 0.1% (wt/wt) or sesamol + α-tocopherol each at 0.01% level to ground beef before irradiation effectively reduced lipid oxidation and VSCs (41). Patterson and Stevenson (23) found that dietary supplementation of α-tocopherol and ascorbic acid to hens reduced the yield of total volatiles. Dietary vitamin E added to turkey diets reduced production of MT, DMS, CS_2 DMDS and some hydrocarbons and aldehydes of raw turkey meat (42). However, Lee et al. (43) found antioxidant combinations (sesamol+α-tocopherol and gallate+α-tocopherol) had very little effect on the development of off-odor and the formation of VSCs due to irradiation in raw turkey meat.

Table IV. Effect of Antioxidants on Irradiation-induced Volatile Sulfur Compounds Production in Turkey Bologna

Antioxidant	H_2S^Z	CS_2	MT	DMS	DMDS	DMTS
None	503 c	8036 a	3759 b	2059 a	5742 b	552 b
Erythorbate	941 a	7165 a	3696 b	1566 b	5888 b	664 b
Nitrite	876 a	8117 a	4696 a	1482 b	7736 a	2375 a
Rosemary	691 b	6999 a	3193 b	1159 b	7475 a	2104 a
$LSD_{0.05}^Y$	147	1417	883	413	1141	983

Vegetable oil (none) or antioxidants in vegetable oil were added as ingredients in the raw meat emulsions for bologna manufacture. The antioxidants tested were 500 ppm sodium erythorbate (erythorbate), 200 ppm sodium nitrite (nitrite), and 0.075% rosemary extract (rosemary). The emulsions were cased, cooked, and irradiated at 3 kGy. Volatile compounds measured the next day, and expressed as square root of peak area count. The numbers were means of four replicates. Adopted from ref (26).
[Z] Numbers in columns followed by the same letter are not significantly different (P<0.05).
[Y] The least significant difference at P<0.05 level.

Du and Ahn (*24*) found addition of sesamol as an ingredient in raw meat mixtures prior to sausage manufacture reduced production of DMDS and DMTS while gallic acid reduced levels of DMDS in irradiated ready-to-eat turkey sausage. Vitamin E and rosemary extract had no effect on production of any VSC although all antioxidants reduced lipid oxidation and color change due to irradiation. Our results showed addition of nitrite, erythorbate, or rosemary extract in raw meat mixtures used for turkey bologna manufacture did not reduce levels of VSCs from irradiated RTE products (*26*). Some of VSCs were even promoted by the addition of the antioxidants (Table IV). To study whether VSCs can be reduced by antioxidant treatment after RTE meats are manufactured. Commercial turkey bologna was dipped into water or different antioxidant solutions for 2 min. The antioxidants were 20 mM of ascorbic acid, vitamin E, and sesamol, and 0.75% of rosemary extract. The samples were then irradiated to 3 kGy at 4°C. VSCs were then measured using SPME-GC-PFPD. Results showed that the antioxidants had no effect on VSCs in the non-irradiated bologna (Table V). Regardless of antioxidants treatment, irradiation induced formation of MT, DMDS and DMTS and reduced production of DMS and CS_2. Sesamol reduced H_2S levels in irradiated samples although irradiation itself had no significant effect on H_2S levels. Overall our results showed that dipping turkey bologna in antioxidants solutions did not reduce the production of VSCs due to irradiation (Table V). It appears antioxidants will have very limited effect on irradiation-induced VSCs. It is unclear why antioxidants reduced the production of some VSCs in raw meat but not in cooked meats. Perhaps, cooking changes the properties of food components and antioxidants, and/or alters the interaction of food components and antioxidants. RTE products may be more sensitive to irradiation in terms of VSCs formation because of increased degradation of proteins/peptides, increasing availability of amino acids. During cooking, cysteine may be formed from cystine due to S-S bond breakdown, and thus formation of VSCs is favored during subsequent irradiation. However, thermal treatments such as those used for processing of RTE meat products also induce production of volatile sulfur compounds. Irradiation may have less impact on actual odor/flavor of RTE meats than raw meats because of the background VSCs already present in RTE meats.

During storage, irradiated meats packaged in air permeable bags had much lower VSCs levels compared to those in oxygen-barrier bags, suggesting that the VSCs are either oxidized or volatilized in air permeable bags. Use of double packaging systems has been proposed by Ahn and his colleagues (*44, 45, 46*) to reduce off-odor in raw meat products. The double bagging consists of an outer air-impermeable bag and an inner air-permeable film bag. After irradiation, the outer bag can be torn to allow escapes and oxidation of VSCs. Most RTE meats are traditionally packaged in air-impermeable film bags. Packaging materials for RTE meats may be so designed to allow VSCs escape from the packages.

Table V. Effect of Antioxidants on Irradiation-induced Volatile Sulfur Compound Production of Turkey Bologna

Dose (kGy)	Antioxidant	H_2S	CS_2	MT	DMS	DMDS	DMTS
0	None	483	4973	247	505	318	107
	Ascorbic acid	419	4224	205	553	163	80
	Vit. E	354	4273	210	521	175	76
	Sesamol	445	4322	255	589	176	78
	Rosemary	370	4172	232	522	174	85
3	None	522	3524	2126	266	4522	418
	Ascorbic acid	759	3145	1410	233	3521	331
	Vit. E	396	2744	1776	233	3958	348
	Sesamol	268	3477	1807	366	4972	562
	Rosemary	407	3202	2103	277	4763	495
$LSD_{0.05}{}^Z$		246	880	404	306	867	132
Dose		NS^Y	***	***	***	***	***
Antioxidant		*	NS	NS	NS	NS	NS
Dose x antioxidant		NS	NS	NS	NS	NS	NS

Diced turkey bologna was dipped in water (none) or different antioxidants solutions before irradiated at 3 kGy. The antioxidants tested were 20 mM of ascorbic acid, vitamin E, and sesamol, or 0.75% of rosemary extract (rosemary). Volatile compounds were measured the next day, and expressed as square root of peak area count. The numbers were means of four replicates

Z The least significant difference at P<0.05 level.

Y NS, *, and *** indicate no significant effect or significant effect at P<0.05 and P<0.001, respectively.

Future Research

Despite years of research on irradiation odor, many questions remain unanswered. There is no systematic descriptive sensory analysis of irradiation-induced odor and aroma in RTE meat products. The description of off-odor has been mostly reported for raw meats. Limited studies have indicated that the off-odor is not always found in low dose irradiation-induced RTE products. The descriptive analysis may be conducted on RTE products irradiated in higher doses, so distinguished odors can be detected.

Although many studies showed the correlation between VSCs and irradiation odor, a detailed determination of the relative contribution of individual compounds to the irradiation odor has not been available. Many techniques may be employed to address the issue. GC-olfactometry (GC-O) is a technique used to characterize the odors of a single component or a complex mixture of volatiles by sniffing the GC effluent after separation. By using GC-O, one can determine the contribution of a given aroma compound as it elutes as an individual peak on the GC. The technique may be used to detect trace amounts of VSCs that have impact on the odor of irradiated meats. Extraction dilution techniques may also be used to rank the key odorants in the order of potency based on evaluation of individual GC peaks by GC-O. The higher the dilution at which a compound can be detected by GC-O, the greater the contribution of that compounds to the odor of the food.

The VSCs in irradiated RTE meats have not been quantified due to the complexity of meats, and the instability, volatility and reactivity of VSCs. So far, VSCs are routinely reported as peak area counts. However, area counts do not always correlate to actual amounts of the compounds. Attempts should be made to accurately quantify the amount of VSCs in irradiated foods, so that the impact of individual VSCs can be determined.

The mechanism for VSC formation induced by irradiation is not completely understood although it is generally believed that free radicals generated from the radiolysis of water are involved. What radical(s) are directly involved in the formation of VSCs is unclear. The mechanism for the formation of VSCs, particularly the thiols, has not been determined. Many antioxidant are very effective in inhibiting lipid oxidation but not effective in reducing VSCs, indicting that the mechanism for VSC formation may be different from lipid oxidation. Understanding the mechanisms will help us to develop measures to reduce production of VSCs and off-odor.

References

1. Zink, D. L. *Emerg. Infect. Dis.* **1997**, *3(4)*, 467-469.
2. Ryser, E. T.; Marth. E. H. *Listeria, listeriosis and Food Safety.* Marcel Dekker: New York, NY, 1999; pp 372, 509.
3. USDA, FSIS. *Recall Information Center.* http://www.fsis.usda.gov/OA/recalls/rec_all.htm. Accessed Jan. 16, 2004.
4. Thayer, D. W.; Lachica, R. V., Huhtanen, C. N.; Wierbicki, E. *Food Technol.* **1983**, *40*, 159-162.
5. Thayer, D. W.; Josephson, E. S.; Brynjolfsson, A.; Giddings, G. G. *Meat Focus Int.* **1996**, *5*, 271-277.
6. Sommers, C. H.; Keser, N.; Fan. X.; Wallace, F. M.; Novak, J. S.; Handel, A. P.; Niemira, B. A. In: *Irradiation of Food and Package: Recent Development*; Komolprasert, V; Morehouse, K., Eds.; American Chemical Society: Washington DC, 2004; pp. 77-89.
7. FDA. *Federal Register.* **2000**, 65(3), 493.

8. Hashim, I. B.; Resurreccion, A. V. A.; Mcwatters, K. H. *J. Food Sci.* **1995**; *60*, 664-666.
9. Reineccius, G. A. J. *Food Sci.* **1979**, *44*, 12-24.
10. Singh, H. 1992. In: *Off-flavors in Foods and Beveranges*; Charalambous, G., Ed.; Elsevier: Amsterdam, Netherlands. 1992; pp 625-664.
11. Ahn, D. U.; Lee, E. J. In: *Irradiation of Food and Package: Recent Development*; Komolprasert, V; Morehouse, K., Eds.; American Chemical Society: Washington DC, 2004; pp. 43-76.
12. Huber, W.; Brasch, A.; Waly, A. *Food Technol.* **1953**, *7*, 109-115.
13. Batzer, O. F.; Pearson, A. M.; Spooner, M. E. *Food Technol.* **1959**, *13*, 501-508.
14. Terrell, R. N.; Smith G. C.; Heiligman, F.; Wierbicki, E.; Carpenter, Z. L. *J. Food Sci.* **1981**, *44*, 215-219.
15. Barbut, S.; Maurer, A. J.; Thayer, D. W. *Poultry Sci.* **1988**, *67*, 1797-1800.
16. Al-Bachir, M.; Mehio, A. *Food Chem.* **2001**, *75*, 169-175.
17. Johnson, A. M.; Reynolds, A. E.; Chen, L.; Resurreccion, A. V. A. *J. Food Proc. Preser.* **2004**. (accepted).
18. Heath, J. L., Owens, S. L.; Tesch, S. *Poultry Sci.* **1990**, *69*, 313-319
19. Ahn, D. U.; Jo, C.; and Olson, D. G. *Meat Sci.* **2000**, *54*, 209-215.
20. Batzer, O. F.; Doty, D. M. *J. Agric. Food Chem.* **1955**, *3*, 64-67.
21. Ahn, D. U. *J. Food Sci.* **2002**, *67*, 2565-2570.
22. Wick, E. L.; Yamanishi, T., Wertheimer, L. C.; Hoff, J. E.; Proctor, B. E.; Goldblith, S. A. *J. Agric. Food Chem.* **1961**, *9*, 289-293.
23. Patterson, R. L. S.; Stevenson, M. H. *British Poultry Sci.* **1995**, *36*, 425-441.
24. Du, M.; Ahn, D. U. *Poultry Sci.* **2002**, *81*, 1251-1256.
25. Fan, X. T.; Sommers, C. H.; Thayer, D. W.; Lehotay, S. J. *J. Agric. Food Chem.* **2002**, *50*, 4257-4261.
26. Fan, X.; Sommers, C. H.; Sokorai, K, J, B. *J. Agric. Food Chem.* **2004**, *52*, 3509-3515.
27. Shankarananarayana, M. L.; Raghavan, B.; Abraham, K. O.; Natarajan, C. P. *CRC Crit. Rev Food Sci. Nutrit.* **1974**, *4*, 395-435.
28. Van Gemert, L. J.; Nettenbreijer, A. H. *Compilation of Odour Threshold Values in Air and Water.* National Institute for Water Supply, Central Institute for Food and Nutrition Research: Voorburg, Netherlands. 1977.
29. Wick, E. L.; Murray, E.; Mizutani, J.; Koshika, M. *Radiation Preservation of Foods, Advances in Chemistry Series*, American Chemical Society, Washington, DC, 1967; pp 12-25.
30. Simic, M. G. In: *Preservation of Food by Ionizing Radiation;* Josephson, E. S.; Peterson, M. S., Eds.; CRC Press, Boca Raton, FL. 1983; Vol. 2, pp. 1-73,
31. Yoo, S. R.; Min, S.; Prakash, A.; Min, D. B. *J. Food Sci.* **2003**, *68*, 1259-1264.

32. Martin, S.; Batzer, O. F.; Landmann, W. A.; Schweigert B. S. *Agric. Food Chem.* **1962**, *10*, 91-93.
33. Wilkening, V. G.; Lal, M.; Arends, M.; Armstrong, D. A. *J. Physic. Chem.* **1968**, *72*, 185-190.
34. Packer, J. E.; Winchester, R. V. *Can. J. Chem.* **1970**, *48*, 417-421.
35. Jo, C.; Ahn, D. U. *J. Food Sci.* **2000**, *65*, 612-616.
36. Merritt, Jr. C. In: *Food Irradiation, Proc. Int. Symp. on Food Irradiation*; Int. Atomic Energy Agency: Vienna 1966; pp. 197-210.
37. Amirav, A.; Jing, H. *Anal. Chem.* **1995**, *67*, 3305-3318.
38. Languorous, S.; Escher, F. E. *J. Food Sci.* **1998**, *63*, 716-720.
39. Hashisaka, A. E.; Einstein, M. A.; Rasco, B. A.; Hungate, F. P.; Dong, F. M. *J. Food Sci.* **1990**, 55, 404-412.
40. Nam, K. C.; Prusa, K. J.; Ahn, D. U. *J. Food Sci.* **2002**, *67*, 2625-2630.
41. Nam, K. C.; Min, B. R.; Park, K. S.; Lee, S. C.; Ahn, D. U. J. *Food Sci.* **2003**, *68*, 1680-1685.
42. Nam, K. C.; Min, B. R.; Yan, H.; Lee, E. J.; Mendonca, A.; Wesley, I.; Ahn, D. U. *Meat Sci.* **2003**, *65*, 513-521.
43. Lee, E. J.; Love, J.; Ahn, D. U. J. *Food Sci.* **2003**, *68*, 1659-1663.
44. Nam, K. C.; Ahn, D. U. *J. Food Sci.* **2002**, *67*, 3252-3257.
44. Nam, K. C.; Ahn, D. U. *Poultry Sci.* **2003**, *82*, 1468-1474.
46. Nam, K. C.; Ahn, D. U. *Meat Sci.* **2003**, *63*, 389-395.

Acknowledgements: The author thanks Doug U. Ahn, Marilyn Schneider, and Robert Gates for reviewing the manuscript, and Kimberly J. B. Sokorai and Robert Richardson for technical assistance.

Author Index

Subject Index

A

Acetate. *See* Glucose-glycine model systems

Acetic acid, reaction time and dry-cured ham content, 80

2-Acetyl-1-pyrroline
glucose/proline mixture, 138–140
stability, 140*t*

2-Acetyltetrahydropyridine, glucose/proline mixture, 138–140

3-Acetylthio-2-alkyl alkanals
olfactophore model, 178, 179*f*
parallel synthesis, 173–174, 177–178
sensory properties, 178, 179*f*
See also Sulfur-containing odorants

Acids
dry-cured ham, 71, 72*t*
effect of adding amino acids on dry-cured ham, 79
effect of sodium chloride content in dry-cured ham, 76, 77*f*
effect of sodium nitrite addition in dry-cured ham, 76, 78*f*, 79
rice bran hydrolyzed vegetable protein (RB–HVP), 94*t*, 95*t*

β-Alanine
browning of sugars with, 160*f*
Maillard browning, 159, 161
minimizing competing Strecker degradations, 159
xylose/β-alanine browning in phosphate and bis-tris buffers, 160*f*
See also Maillard browning

Alcohol
halal requirements, 60–61
kosher versus halal, 65

Aldehydes
dry-cured ham, 71, 72*t*
effect of sodium chloride content in dry-cured ham, 76, 77*f*
effect of sodium nitrite addition in dry-cured ham, 76, 78*f*, 79
poultry flavor, 11
reaction time and dry-cured ham content, 80
rice bran hydrolyzed vegetable protein (RB–HVP), 94*t*

Alkylphenols, sheepmeat, 13

Allergens
definition, 51*t*
labeling issues, 51

Amadori compounds of cysteine
browning in cysteine-sugar systems, 123*f*
continuous process for production of D-xylulose-L-cysteine, 120
effect of pH on decomposition of D-xylulose-L-cysteine, 127*f*
experimental procedures, 118–120
formation in cysteine-amino-acid-glucose model systems, 125*t*
general procedure for reactions of cysteine with sugars, 119
generation of volatile sulfur compounds, 123–124
high performance liquid chromatographic (HPLC) analysis, 118
isolation of ketose-L-cysteine, 119–120
liquid chromatographic analysis of cysteine and xylose during heating, 121
materials, 118
maximizing flavor yields in cysteine-sugar systems, 126–128

225